In Pursuit of Moby-Dick

Joseph S. Catalano

In Pursuit of Moby-Dick

Of Whales and Their Gods

palgrave
macmillan

Joseph S. Catalano
New York, NY, USA

ISBN 978-3-031-40356-9 ISBN 978-3-031-40357-6 (eBook)
https://doi.org/10.1007/978-3-031-40357-6

This Palgrave Macmillan imprint is published by the registered company Springer Nature Switzerland AG.
The registered company address is: Gewerbestrasse 11, 6330 Cham, Switzerland
Paper in this product is recyclable.

For my brother Charles and his family.

Preface: I Will Write

The first reference to my new long book about whales and whaling was, I think, given by Richard Henry Dana Jr., who had recently published *Two Years Before the Mast*. I saved my reply:

> About the "whaling book"—I am half way in the work, & am very glad that your suggestion so jumps with mine. It will be a strange sort of a book, tho', I fear; blubber is blubber you know; tho' you may get oil out of it, the poetry runs as hard as sap from a frozen maple tree;—& to cook the think up, one must needs throw in a little fancy, one ungainly as the gambols of the whales themselves. Yet I mean to give the truth of the thing, spite of this.[1]

Yes, dear reader, that is the truth of my book, *Moby*-Dick. What truth? Why simply this, that the poetry comes from the marriage of whale's blubber and those that seek deep within it. This is the clue to the proper reading of my *Moby-Dick*. Yes, to anticipate, there is Ahab's mad chase after the great white sperm whale, Moby Dick, and a great deal about myself as Ishmael and my loyal cannibal harpooner, Queequeg? They are all present and important; but, do not be let astray, either separately or together, their personalities and adventures will never unite this tale into one great narrative. Only the life of whales and the adventure of whalers can give my book its true unity. This is my general claim in all that follows; but, I ask you, dear reader, to absorb the weight of this claim, for the largest whales can stand comparison with the biggest dinosaurs, and whales are still with

[1] Hershel Parker, *Herman Melville: A Biography Volume 1, 1819–1851*, Page 724.

us—they have been with us for thousands of years. You know all of this? Still, let me go on, for we all need to find the right words in all that is important for us. Here is how it all began; the beginning was as strange as the voyage itself.

New York, NY, USA

Joseph S. Catalano

Acknowledgments

Eileen Srebenik guided the book from beginning to end and it would not exist without her initial and continued interest. Esther Rani was the production editor, and V. VinodhKumar helped with the final stages. Everyone else at Palgrave has been efficient and courteous. My wife, Marisa, helped through many stages and helped with the Index.

Contents

About the Author

Joseph S. Catalano Professor Emeritus of Kean College of New Jersey.
Publications:

A. Seven Books: (1) *A Commentary on Jean-Paul Sartre"s ""Being and Nothingness."* New York: Harper & Row, 1974; Rev. ed. Chicago: University of Chicago Press, 1980; Midway Reprint: University of Chicago Press, 1985. (2) *A Commentary on Jean-Paul Sartre"s ""Critique of Dialectical Reason."*" Chicago: University of Chicago Press, 1986. (3) *Good Faith and Other Essays: Perspectives on a Sartrean Ethics.* Lanham: Rowman & Littlefield, 1996. (4) *Thinking Matter: Consciousness from Aristotle to Putnam and Sartre.* Routledge: New York and London, 2000. (5) *Reading Sartre: An Invitation.* Cambridge, Cambridge University Press, 2010. (6*) The Saint and the Atheist: Thomas Aquinas and Jean-Paul Sartre,* University of Chicago Press, 2021. (7) *The Family Idiot: Abridged with an Essay by Joseph S. Catalano*, University of Chicago Press, January 2023.

B. In Progress: *One with a Book: The Magic and Mystery of the Written Word.*

C. About 30 or so articles, 15 or so reviews, 10 or so articles in Anthologies, most recent, two for Routledge, *Consciousness and the Great Philosophers*, 2017, and *The Meaning of Life and the Great Philosophers, 2018*, both on Sartre.

D. Honors, National Endowment for the Humanities: Year grant for independent study 1983–1984; three summer grants; two year post-doctoral grant, Princeton, 1984–1986. Papers delivered, On the average of one a year from about 1972–2003.

CHAPTER 1

No Need to Rush

Abstract I take my readers on a personal tour through Melville's great novel, *Moby-Dick*. I do not assume any knowledge about the book as I share more than twenty years of reading.

In reply to a letter by Richard Henry Dana Jr., author of *Two Years Before the Mast*, published about ten years before *Moby*-Dick, Melville gives us the first known reference to his forthcoming book.

> About the "whaling book"—I am half way in the work, & and am very glad that your suggestion so jumps with mine. It will be a strange sort of a book, tho', I fear; blubber is blubber you know; tho' you may get oil out of it, the poetry runs as hard as sap from a frozen maple tree;—& to cook the think up, one must needs throw in a little fancy, one ungainly as the gambols of the whales themselves. Yet I mean to give the truth of the thing, spite of this.

"I mean to give the truth of the thing, spite of this." What truth? Why simply this, that the poetry comes from the blubber? Yes, and from the truth about the blubber. This is the clue to the proper reading of *Moby-Dick*. And, it is more than a clue. On every page we witness the marriage of poetry and blubber; on every page we witness truth being born from their marriage. If you separate these three—the poetry from the blubber and both from truth—you then hold in your hand not one great book but three novellas loosely pasted together.

J. S. Catalano, *In Pursuit of Moby-Dick*,
https://doi.org/10.1007/978-3-031-40357-6_1

The vastness and complexity of the literal meaning of *Moby-Dick*—the way poetical, religious, and philosophical reflections are intertwined with the flesh and skeleton of whales and the details of American whaling ships—has confused not only its early critics and readers but many of those of today.

It was not Ahab that drew me back to reading *Moby-Dick* on and off, at times only a chapter here and there, but, from beginning to end, at least ten to fifteen times. Nor was it my philosophical career, for aesthetics was never my forte. It was the *story itself!* Poetry and truth I could find elsewhere, but not as they are one with whales. These leviathans, traveling through all the oceans of the earth as easily as we walk our pathways and streets, spout not merely vapor but a story of the planet Earth itself.

> There Leviathan
> Hugest of living creatures, in the deep
> Stretch'd like a promontory sleeps or swims,
> And seems a moving land; and at his gills
> Draws in, and at his breath spouts out a sea

These words from Milton's *Paradise Lost* are on the title page of the true first edition of *Moby-Dick*; and, further, its title is simply, *The Whale.* As I collector, I have always wanted to own that book, but it was beyond my means, for only 500 copies were printed in England by Richard Bentley in October 1851. Its production was strange and somewhat contradictory; for, while it came out in three impressive-looking volumes, it was incomplete. Bentley trimmed Melville's classic by thirty-five passages that he thought offensive. He also put the two preliminary sections of the book, "Etymology," and "Extracts," at the end; and he left out the last page, "Epilogue," describing Ishmael's escape from the doomed ship, Pequod, with the result that the British readers and critics wondered how this tale of more than 600 pages could have been narrated.

A short time after its publication in England, Melville delivered a complete and properly organized manuscript to Fletcher Harper, one of the Harper brothers, of Harper & Brothers in America. There was one important change. Harper had recently published Henry T. Cheever's, *The Whale and His Captors*, and, perhaps, anticipating complaints, Melville altered the title to *Moby Dick; or, The Whale.* Notice the semi-colon; it does not point to a redundant expression: Moby Dick=The Whale. Rather, the title means Moby Dick, *and* the Leviathan, The Whale. It is the true complete title of the book in the first American edition, although it is seldom repeated. It is more convenient to refer simply to *Moby-Dick*, and I will frequently follow that custom, although I cannot wonder if our reading would have been guided more properly by the original title, *The Whale.*

Do not be a reading-Ahab hastening toward that fast-passed concluding chase; if that is what you desire, see the movie, which is fun, and has its place.

Keywords White Whale • Ahab • Pequod, • Merchant Service • Queequeg

"Rest," they told me, "there is no need to rush, and if there is anything you need, let us know." Well, that was pleasant to hear, and yes I rested. It is more than two weeks since that momentous trip aboard the Pequod with Ahab and his wild search for the great white whale Moby Dick. There was a lot to clear up. I do not wish to go into all that just yet; but now everything seems to be in order. Still, they are concerned. I understand. And, to be honest to you, I have my own motives for writing. It was a long voyage during which I fell in love with whales and the men who hunt them. If you knew me, you would not be surprised that I could not stop taking notes. In truth, they are giving me an opportunity to put my reflections in order, and I will do just that. Of course, I must keep their needs in mind—how I am to handle *that* affair I do not know. Perhaps, writing about it will suggest a path to follow.

I will tell you everything! Why not! I had then no clear notion of the finished manuscript. Would I ask for help, for some suggestions before I submit it? But again, if you knew me, you would know that once I have made up my mind, I never do listen to advice, even when I ask for it. On the way to my finished product, you know that there are always one or two drafts before the completed manuscript; and here I am open—or am I?

From the very beginning it was clear to me that this writing was going to be completely different from anything else that I have written, and that I needed a method. I had before me mountains of notes and reflections, and I get dizzy as I turn one page or torn sheet after another. What about my poor readers? A little at a time Melville—I never refer to myself as

Herman. There is something hopeful in finishing a chapter, and I wish to give my readers a great deal of hope; for before I started I knew it was to be a long and complex tale. Still, I will be clear about what is clear; and much of it is clear, even bright—yes white, very white—like that whale, Moby Dick. Complexity does not always arise from difficult situations pregnant with allegory and myth, but from what is evident, what shines bright before us. Frequently, the evidence before us is so clear that we blink, and within the times of those blinks, we have either altered our lives or decided the evidence was a mirage. I tell you the great truths about whales and the men who hunt them are clear as light, despite whatever darkness was gradually born within Ahab when he was aboard the Pequod.

And, I will put simple titles to my chapters. These are important, for I understand a title to be like an arrow, no better like a vector, pointing the way to whatever complexity the chapter might have—the vector embodies the weight of a specific origin, my decision to go whaling. How many chapters will I have? Forty, I think. Yes, I can almost see those forty chapters, even though not one is yet written. Aside from writing about my voyage, you may be surprised to learn that I almost did not go. I am a seasoned sailor with merchant service, but those relatively short and safe trips merely highlighted the uniqueness of that voyage on the Pequod. If I had any doubts, retired captain Peleg, one of the owners set me straight, "Merchant service be damned. Talk not that lingo to me." Perhaps he thought that I expected to have a privileged position, but I was content to be an ordinary seaman. Or, maybe he thought I was expecting an extra share in the profits because of my Merchant service. He clarified that issue; but he showed himself to be generous compared to his partner, captain Bildad. Everyone on board the ship received a percentage of the profits from the sale of the whales, when the long voyage of two, three, or even four years is over. I did not know that this percentage is termed a "lay," and while captain Bildad, suggested only "the seven hundred and seventy-seventh lay," the more reasonable captain Peleg gave me the three-hundredth lay. In either case, I was not to plan on retiring in luxury from this trip. It was a completely different affair when I later introduced them to my new friend, the harpooner Queequeg who was not at all shy and knew his worth:

Queequeg

> Cap'ain, you see him small drop tar on water dere? You see him? well spose him one whale eye, well den! ... and taking sharp aim at it, he darted the iron right over old Bildad's broad brim, clean across the ship's decks, and struck the glistening tar spot out of sight.[1]

Peleg gave Queequeg a ninetieth lay, and remarked, "and that's more than ever was given a harpooner yet out of Nantucket."

But even as a whaling voyage, I was soon to learn that in choosing the Pequod I had fixed a destiny for myself that even a seasoned whaler might never experience. I am thinking now of those rare but instructive meetings between two whaling ships, "gams" they called them. Perhaps, you know the term, but it was new to me. During these gams I gradually learned that what was occurring aboard the Pequod was extraordinary, even among the ordinary extraordinary happenings on any whaling voyage. There were nine gams, each unique. I will describe each one; I know that you will be interested—already I suspect more than forty chapters, oh well, what will be will be.

Have you heard about Captain Ahab? Good! And if not, I will tell you. I can here anticipate my descriptions by observing that partly the magic aboard the Pequod arose from Ahab, a true whaling man, merely catching his breath on shore before being ready to ship out again on a long three-year whaling voyage. We all knew that Ahab had his leg torn off by that great white sperm whale that had been given the name Moby Dick. Are you surprised that whales have names? There is Timor Jack, New Zealand Tom, and Don Miguel. Other captains beside Ahab went out of their way to kill these great whales, and so I must tell my tale in such a way that you will understand the true distinctiveness of Ahab's search. Ahab's leg had mended well and it had been neatly replaced with a proper fitting whale-bone leg, below the knee. The two principal owners of the ship, retired captains Peleg and Bildad, were not surprised when Ahab wanted to ship

[1] Ishmael has gone on another long two- or three-year voyage, and the curious might wish to know that he will be visiting Queequeg, who, since his father died, is now King of his realm. I share with you my thoughts before the publication of the book, but it is now here besides me. I think Ishmael will be pleased with the production, although I wish it could have been done on better paper. The publishers sent me ten copies, and I have distributed them, keeping three for when Ishmael returns. I am, of course, using the published book, to review with you my thoughts while writing it. This first quote is from chapter 18, "His Mark."

out again, despite his having a young wife, "resigned" they say, and a young child.

Ahab was wedded to whaling, and yet I met others, more tightly bonded. I recall Bulkington. (You say to yourself, I wander. Let it be.) I first saw him in the Spouter-Inn before I had picked out the Pequod as my whaling ship. He and some of his sailor friends had just returned from a long whaling voyage and were enjoying a relaxing drink. He was reserved but clearly liked by his companions. A mere few days later I saw him again on the Pequod. We had just started out and we were steering clear of the magnetic draw of the lee shore before a ship reaches open seas. Bulkington was standing at the helm. I was awed. I went below and scribbled one of my notes. Here is what I wrote: "I looked with sympathetic awe and fearfulness upon the man, who in midwinter just landed from a four years' dangerous voyage, could so unrestingly push off again for still another tempestuous term." And, of course, I added a reflection. Here it is on the same torn piece of paper (I must put them in order if I am going to write about this tale):

> But as in landlessness alone resides the highest truth, shoreless, indefinite as God—so better is it to perish in that howling infinite, than be ingloriously dashed upon the lee .... Terrors of the terrible! is all this agony so vain? Take heart, take heart, O Bulkington! Bear thee grimily, demigod! Up from the spray of thy ocean-perishing—straight up, leaps thy apotheosis. (Chapter 23, "The Lee Shore")

Melville, if you are going to write like that what is to become of your tale? Well, I have a philosophic spirit and I brought some books along. I choose to go whaling, and philosophy, poetry, and truth, as they are one with the whales that travel the oceans of the world—that is what I wanted to capture. I recall that when there was a need to balance the Pequod, which had a sperm whale tied on the other side, the harpooners lowered their dignity to catch a right whale to balance the ship. I had just finished browsing through a short history of philosophy, reviewing my earlier studies on Locke and Kant:

> So, when on one side you hoist Locke's head, you go over that way; but now, on the other side hoist in Kant's you come back again; but in very poor plight. Thus some minds keep trimming boat. Oh, ye foolish! throw all these thunder-heads overboard, and then you will float light and right. (Chapter 73, Stubb and Flask Kill a Right Whale)

Philosophy is nothing if it does not aid you to think for yourself. We teach our students philosophy too late in life; it has to be learned early as a natural part of their spontaneous questioning about the world. But, we alien being, adults, think we have all the answers. Learning about philosophy should be an aid to our own thinking about the world, whether it is the distinctive empiricism of the English philosopher John Locke or the more speculative German philosopher Immanuel Kant we should turn to them because we see them clarifying and enlightening our own thoughts. Otherwise, one has only a bit of this and a bit of that, and no real self-enlightenment.

You must excuse my going on like this, but it does make me feel better—I keep that difficult task of clarifying what happened aboard the Pequod at a distance. Besides, I cannot take a step without wondering where my feet are taking me or how the movement of my entire body is possible through my steps. But even with whales, poetry, and philosophy, I am young and strong enough so that I stand tall and also do my duty to the ship—most of the time. Still, I tell you that I would give this advice to the owners of whaling ships: "Let me in this place movingly admonish you … Beware of enlisting any lad of lean brow and hollow eye … this sunken-eyed Platonist will tow you ten wakes round the world, and not make you one pint the sperm richer (35, the Mast Head)." I tell you that I wavered between Plato and his student Aristotle with those huge waves that almost sunk us. At times with Plato I thought this world a mirage merely reflecting a nobler world of pure spirit in which are the archetypes of all that we hold true, and which as we know Augustine, several hundred years later, place in the mind of God. But, then, the Pequod would go down some mountain of a wave, fresh salty water awakening me to the world about me, the ship rising safely up another mountain of a wave, and with Aristotle I would bless the wonder of the world about me, and like Aquinas a thousand years later, be joyful of both worlds without melting one into the other.

A Gam

I had mentioned that the slippery magic aboard the Pequod arose from Ahab, but if so it sunk deep within the very architecture of the ship, its "Try-Works" that announce the appearance of an American whaling ship, and, from this architecture into the mates, carpenter, blacksmith, cabin boy, and into me, Melville, who called myself Ishmael on this voyage, and into every other sailor. Why did it happen and how? Was it all due to Ahab? Yes, but then no! I mentioned the gams, and there was one that was particularly instructive and enlightening concerning Ahab and Moby Dick. Most of our meetings were with other American whaling ships—they are now the most numerous—but the one I now wish to tell you about was with a British ship, the Samuel Enderby of London. Usually, in the gams, Ahab does not board the visiting ship, merely asks about whether they have seen the white whale, Moby Dick; but this time the captain of the ship, Boomer, had his arm chewed off by Moby Dick, and Ahab went aboard the Enderby—I, Queequeg, and the very strange, Fedallah, to whom I will return, were with him. Never did two men react so differently to similar events. For Boomer it was an exciting incident, not only for himself but for his surgeon, Dr. Bunger. I will not detail the conversations, but picture the gathering: Boomer joyfully describing his encounter with Moby Dick, while his surgeon, Dr. Bunger, would gleefully interrupt, describing how he handled the wound and gave Boomer a new whale-bone arm. Ahab was going crazy. Once he discovered that Boomer had not killed Moby Dick—Ahab was relieved to hear this—he could not wait to leave. But Boomer noticed Ahab's impatience, and when he becomes aware of Ahab's vengeful search after Moby Dick, he whispered to Fedallah "Is your Captain crazy" (chapter 100, The Pequod meets the Samuel Enderby of London)?

Captain Boomer did not know that Fedallah was as weird as Ahab. Ahab had sneaked him and a few others aboard, to serve him and him alone, anticipating that he would need special help to kill Moby Dick. If the owners had known, they would have stopped the voyage before it began. Indeed, Fedallah was a man of omens and portents. There were portents and omens throughout this voyage; and yet all was clear, until the end, all had natural explanations and even Ahab tended not to believe in them. I recall that he said, "Omen? omen?— the dictionary! If the gods think to speak outright to man, they will honorable speak outright (133, The Chase—The First Day)." And yet, he could not dispel their presence from his own mind.

Ahab kept confiding in me; I was a good listener, and I expressed no opinions of my own. But I saw that something was changing within him when toward the very end, he said, "The things called omens! … Oh! how valently I seek to drive out of others' hearts what's clinched so fast in mine!—The Parsee—The Parsee! Gone, gone? … *I'll, I'll* solve it, though (134, The Chase, The Second Day)."

Ahab as Captain and Ahab as Ahab

Well, he did solve it. How? I mentioned that the owners had their own plans for what I was to write, and that they were paying me for the writing. I had to keep faith to my conscience, and yet keep their needs in mind. I think that I did it; but not by writing one tale but two tales in one. Still, with all this ambiguity, the voyage was very successful. Ahab was intelligent, and he knew his men. He let them catch their sperm whales, as long as he was heading in the direction in which he thought he would discover the white whale Moby Dick. Yielding to my philosophic bent, I make a distinction between Ahab in his mad pursuit after Moby Dick and Ahab as a competent captain of the whaling ship Pequod. Most of the time, indeed, until the very ending of the voyage, Ahab let the mates and harpooners do their jobs, and they did them well.

I must not let mad Ahab cloud the everyday wonder of whales and the men who pursue them. Oh, I know that some would say that no one should kill whales, and yet every part of the sperm whale was used; there was no waste. But, actually, these killings of sperm whales could be more accurately compared to medieval knights jousting than to war, or to the American Indians who killed a buffalo because every part of the animal was used to fulfill the needs. Then too, whales cannot be so easily cornered for a mass killing as were buffalos by the insane profit hunters, "Forty men in one ship hunting the Sperm Whale for forty eight months think they have done extremely will and thank God, if at last they carry home the oil of forty fish (105, Does the Whale Diminish?)."

Thus, although Ahab's personal pursuit after Moby Dick was insane and evil, it was not so crazy that it could not be somewhat placed as part of the natural whaling experience. The harpooners, Queequeg, Tashtego, and Daggo, were all temporarily taken in by Ahab's monomania, for they saw in it the glory of the chase and not the evil in Ahab's intention of killing a whale for the insane purpose of personal vengeance, which led to that insane and inane last chase—a chase to think about in my two-fold tale.

Yes, the ending of that voyage is not easy to understand, and still, I, Melville wrote it! And yet writing is not an innocent affair. Writing about what happens is not what happens. Does it point to what happens? Does it give new life to what happens? There are books and there are things. Are things their own books and are books things? There was a giant white whale named Moby Dick and suppose I write a book named Moby-Dick? Yes, but the book must be published, and the whale just swims on its own. Still, once published, the book is *there* as whales are *there*. So, now that the book exists with whales, it let us now begin to enter both.

Reference

Parker, Hershel, *Herman Melville: A Biography Volume 1, 1819–1851,* Page 724.

CHAPTER 2

The Book Itself

Abstract This introductory section belongs where it is placed by Melville, at the beginning of the book. It prepares us for the narrative that is to follow; and, yet, it is not really part of the story. Its structure is stranger than its content. It is preceded by humorous parenthetical subheadings and remarks. Is Melville teasing us; or does he have something deeper in mind? For the present, let us accept the most obvious explanation, namely, that he is giving us some factual information about whales and whaling, and yet he does not wish us to examine each and every reference as if it were the "gospel" truth.

Keywords Herman Melville • Nathaniel Hawthorne • Moby Dick • Whales • Extracts • Sperm Whales • Etymology

I have a copy of my new book before me, and here is how it begins:

MOBY- DICK;

or,

T H E W H AL E.

BY

J. S. Catalano, *In Pursuit of Moby-Dick*,
https://doi.org/10.1007/978-3-031-40357-6_2

HERMAN MELVILLE,
AUTHOR OF

"TYPEE," "OMOO," MARDI," "WHITE-JACKET

NEW YORK:
HARPER & BROTHERS, PUBLISHERS
LONDON: RICHARD BENTLEY

1851

IN TOKEN

OF MY ADMIRATION FOR HIS GENIUS,

This Book is Inscribed

To

NATHANIEL HAWTHORNE

For now, I would only bring to attention this dedication. Hawthorne is noted for his "twice-told tales," whatever these may be. And, whatever my tale may be, it is also twice-told. For that, I ask patience.

I invite you to take a bird's-eye view of my book, and I will keep in mind that you probably do not own a copy. There are 135 chapters. I claim—indeed, I will prove—*that, at most 40 of the 135 chapters are concerned with Ahab's monomania as distinct from Ahab as any captain of an American whaling ship*! True, Ahab is captain of the Pequod for the entire voyage; and yet, he controlled his spirit of vengeance for most of the voyage. Ahab had revealed his purpose as early as chapter 36, "The Quarter-Deck," during which his first mate, Starbuck, objected, "I came here to hunt whale, not my commander's vengeance. How many barrels will thy vengeance yield thee even if thou gettest it, Captain Ahab?" Ahab was afraid he had gone too far too early in revealing his private intentions. He now decides to keep his private hate in check, and let the Pequod sail normally and catch other whales, until Moby Dick is actually cited:

> Having impulsively, it is probable, and perhaps somewhat prematurely revealed the prime but private purpose of the Pequod's voyage, Ahab was now entirely conscious that, in so doing, he had indirectly laid himself to the unanswerable charge of usurpation; and that with perfect impunity, both moral and legal, his crew if so disposed, and to that end competent, could refuse all further obedience to him, and even violently wrest from him the command. (Chapter 46, "Surmises")

Thus, *for most of the voyage, indeed, until, chapter 128, "The Pequod Meets the Rachel," Ahab keeps his monomania in check.* In that episode, Captain Gardner asks Ahab to help him find his son who is lost at sea. But, Gardner has just sighted Moby Dick, and Ahab stresses his rejection "in a voice that prolongingly molded every word—'Captain Gardner, I will not do it.'" Until that episode, Ahab's dramatics never lead to a clear act of injustice.

This is the basis for my distinction between Ahab as Captain and Ahab as Ahab. That is to say, if you are to read my book, accept that, after my chapter 22, "Merry Christmas," concerned with the loading of the ship, our tale takes place on an American whaling vessel out for a three-year whaling adventure. And that ship must have a captain, and that captain will have idiosyncrasies, which is to say, he will be a particular captain and not a stereotype. Nevertheless, the ship is out to make profit, and that the owners would not have tolerated a captain sailing mainly to satisfy his personal revenge: "Had any one of his old acquaintances on shore but half dreamed of what was lurking in him then, how soon would their aghast and righteous souls have wrenched the ship from such a fiendish man!" (chapter 41, Moby Dick).

On the other hand, I grant the presence of his monomania, before chapter 128; but, it does not amount to more than forty or so chapters—yes, indeed, it became a very big book. If we were to imaginatively separate these forty chapters, and, if for completeness, include the first fifteen chapters about Ishmael and his meeting with Queequeg, we will then have a short book of fifty-five chapters, a swift-paced book, albeit, with numerous reflections on life, madness, evil, and the strange occupation of pursuing whales in general and Moby Dick in particular. Indeed, this fifty-five-chapter book would be generous in what it offers the reader; it would be a fascinating book. On the other hand, *we then have eighty-five chapters dangling in limbo, chapters that center on whales and whaling.*

Furthermore, lest one assert that as my narrative grows, Ahab becomes more dominant, I draw attention to an almost continuous discussion of whales and whaling from chapters 55 to 90, during which less than 10% of the narrative concerns Ahab's madness. Let us consider some of the titles of these chapters: 56, "Less Erroneous Pictures of Whales," 57, "Of Whales in Paint, in Teeth, etc.," 61, "Stubb Kills a Whale," 64, "Stubb's Supper," 74, "The Sperm Whale's Head," 82, "The Honor and Glory of Whaling," 86, "The Tail," 89, "Fast Fish and Loose Fish," 90, "Heads or

Tails." Indeed, my discussion of whales and American whaling occupies more than a half of his entire tale.

Thus, aside from my own comments about my book—for an author may be mistaken about the meaning of his or her work—the book itself bears out the truth of my words of melding adventure, poetry, and truth with blubber. Almost every chapter illustrates this truth. I give here only three examples. Ishmael begins chapter 60, "The Line," with these words:

> With reference to the whaling scene shortly to be described, as well as for the better understanding of all similar scenes elsewhere presented, I have here to speak of the magical, sometimes horrible whale-line.

This very strong but narrow line, merely "two third of an inch in thickness," is attached to the harpoons. It is both necessary and dangerous; for, if it does not uncoil properly, it can cut off an arm, or, even decapitates one—as my Ishmael, *in his writing of this tale*, will let it happen to Ahab in chapter 135. And yet, whatever Ishmael's intentions in writing, this truth remains:

> All men live enveloped in whale-lines. All are born with halters round their necks; but it is only when caught in the swift, sudden turn of death, that mortals realize the silent, subtle, ever-present perils of life. And if you be a philosopher, though seated in the whale-boat, you would not at heart feel one whit more of terror, though seated before your evening fire with a poker, and not a harpoon, by your side. (Monkey-Rope)

Also, consider chapter 72, "The Monkey-Rope." This rope holds the person cutting the blubber from a captured whale being held on the side of the boat—a large whale is much too heavy to be brought on deck. In this case, Queequeg is doing the cutting; and Ishmael is holding the rope that supports him. My Ishmael reflects:

> So strongly and metaphysically did I conceive of my situation then, that while earnestly watching his motions, I seemed distinctly to perceive that my own individuality was merged in a joint stock company of two .... I saw that this situation of mine was the precise situation of every mortal that breathes; only, in most cases, he, one way or other, has this Siamese connexion with a plurality of other mortals.

And, finally, toward the end of chapter 98, "Stowing Down and Cleaning Up," a chapter in which we read about the endless cleaning of the ship, the removal of all blood and blubber from the decks, until another whale is sighted and a sailor shouts:

> "There she blow!" and away they fly to fight another whale, and go through the whole weary thing again. Oh! my friends, but this is man-killing! Yet this is life. For hardly have we mortals by long toilings extracted from this world's vast bulk its small but valuable sperm; and then, with weary patience, cleansed ourselves from its defilements, and learned to live here in clean tabernacles of the soul; hardly is this done, when—*There she blows!*—the ghost is spouted up, and away we sail to fight some other world, and go through young life's old routine again.

These thoughts do not need allegorical interpretations. They contain their own analogies, and these are, for the most part, clear. True, each reader brings to a text a different history and a text thus unveils itself in varying degrees; but that is true of all writing, whether it be about whales, sports, or world poverty. The important point, however, is that neither the whale, Moby Dick, nor the book, *Moby-Dick*, is "a monstrous fable, or still worse and more detestable, a hideous and intolerable allegory" (chapter 45, "The Affidavit"). Both the whale and the book are true, albeit as this truth is one with my fictional adventure of whales and whaling. Is this too much to ask from a book? It is too much to ask from an ordinary book, but not, I hope, from my *Moby-Dick*.

In brief, then, neither Ishmael's interesting persona nor Ahab's fixation on Moby Dick unifies this great book. The unity arises from the way these two tales merge into one larger than either of them separately, namely, the vast epic about whales and the details of whaling. But, to grasp this unity, you have to read leisurely, letting the poetry and truth gradually seep within you, while you live alongside the exploits of these American whalers. As Ishmael I help, for I'll never tire of singing the praises of American whalers, or of whales themselves. In chapters 26 and 27, "Knights and Squires," I refer to the three mates of an American whaling ship as "knights" and to the harpooners, who are associated with them as "squires." They are singular in their chosen dangerous occupation; but they are, at heart, ordinary people:

> But this august dignity I treat of, is not the dignity of kings and robes, but that dignity which has not robed investiture. Thou shalt see it shining in the arm that wields a pick or drives a spike; that democratic dignity which, on all hands, radiates without end for God; Himself! (Chapter 26, "Knights and Squires")

Owning the Whale

On land, these brave sailors are shy and restrained, even when eating together in a pub after a voyage (chapter 5, "Breakfast"). Engaged in whaling, they are unconquerable hunters. Thus, when one hunts whales one is a member of 'Perseus, St. George, Hercules, Jonah, and Vishmoo! there's a member-role for you! What club but the whalesman's can head off like that?' (chapter 82, "The Honor and Glory of Whaling").

Lessons and reflections upon life fill my *Moby-Dick*; but, to repeat, *this tale is also—indeed, it is first and foremost—a great adventure story.* In chapter after chapter, I give details of whaling boats, their crews, the chase and capture of sperm whales, the cutting of their blubber and the careful collecting of the precious oil, and always—as if these 135 chapters reflected the 150 chapters of the Psalms of the Bible—we are given reflections on life.

It has been said, and I think correctly, that only by describing a singular human action does one also exhibit the universal in human life. This can be done very strikingly in great literature. The adventure of whaling is thus not incidental to this narrative; it is that from which and by which we also receive all the poetry and all the truth. It is that adventure which, I hope, will keep people reading and re-reading this book for years to come. And, thus, I recommend that to read *Moby-Dick* properly, we all join hands with me and "own the whale": "For unless you own the whale, you are but a provincial and sentimentalist in Truth" (*Moby-Dick*, 76). Thus, do not be a reading-Ahab, hastening toward the fast-paced chase of my last three chapters—there was, as you will see, special reason for my writing them—for, if you do, you will lose sight of the leviathan, spouting poetry and truth as it roams the oceans of our planet:

> There Leviathan
> Hugest of living creatures, in the deep
> Stretch'd like a promontory sleeps or swims,
> And seems a moving land; and at his gills
> Draws in, and at his breath spouts out a sea

These words from Milton's *Paradise Lost* are on the title page of the true first edition of my book, published in England. The title of that first British edition is simply, *The Whale.* Only 500 copies were printed in England by Richard Bentley in October 1851. Its production was strange and somewhat contradictory; for, while it came out in three impressive-looking volumes, it was incomplete. Bentley trimmed my book by thirty-five passages that he thought offensive. He also put the two preliminary sections of the book, "Etymology," and "Extracts," at the end; and, he left out the last page, "Epilogue," describing Ishmael's escape from the doomed ship, Pequod, with the result that the British readers and critics wondered how this tale of more than 600 pages could have been narrated.

A short time after its publication in England, I delivered a complete and properly organized manuscript to Fletcher Harper—one of the Harper brothers—of Harper & Brothers in America. There was one important change. Harper had recently published Henry T. Cheever's, *The Whale and His Captors*, and, anticipating complaints that I was duplicating Cheever's efforts, I decided to change the title of my book from *The* Whale, to *Moby-Dick; or, The Whale.* Notice the semi-colon; it is not meant to be redundant. "*The Whale*" in the title does not merely refer back to the one whale, Moby Dick, but it means all whales, and this is evident from the book itself. When talking to my relatives or friends, I also refer to my book as *Moby*-Dick, and yet, I often prefer the title of that strange original London edition, ***The Whale***. But now, let us enter deeper into the book itself, and sail again alongside those giants of the deep, the whales. I repeat, the complete American edition is here beside me, and to aid in the strange ending of this book that I wrote, I will now review it with you in more detail. I will keep the references to myself as the Ishmael who wrote it, while also now and then retaining the reference to myself as Melville. And yet again, a*ll this is now secondary to discovering the true meaning of what I have written; for, you must recall with me that those who paid for my writing had their own aims! And so let us begin again!*

CHAPTER 3

Etymology and Extracts

Abstract This second introductory section belongs where it is placed by Melville, at the beginning of the book. It prepares us for the narrative that is to follow; and, yet, it is not really part of the story. Its structure is stranger than its content. It is preceded by humorous parenthetical subheadings and remarks. Is Melville teasing us, or does he have something deeper in mind? For the present, let us accept the most obvious explanation, namely, that he is giving us some factual information about whales and whaling, and yet he does not wish us to examine each and every reference as if it were the "gospel" truth.

Keywords Etymology • Sperm Whale • Right Whale • Nantucket • Herman Melville

Consumptive Usher and Sub-Sub-Librarian 1

I had fun writing these two introductory sections, and I want to share this with you. I did not intend that my readers examine in any great detail either the etymological definitions or the eighty-odd extracts. Their

(Supplied by a Late Consumptive Usher to a Grammar School, and by a Sub-Sub-Librarian)

J. S. Catalano, *In Pursuit of Moby-Dick*,
https://doi.org/10.1007/978-3-031-40357-6_3

general purpose is to impress upon my readers the greatness of whales and whaling as well as their continual presence within our history from biblical times to the present. We tend to take for granted some of the most spectacular things of nature. I suspect that one who is frequently in the presence of whales never takes for granted these remarkable beings that inhabit the oceans of our planet.

True, our flesh and blood whales do not travel among the stars, but they do travel among the oceans of the earth as easily as we walk the streets and pathways of our land. As they come up for air, these leviathans spout the truth of our planet, and it is with such wonder that the whalers behold them. "There she blows," the sailors shout when they spot a whale. "Had the trump of judgment blown, they could not have quivered more; yet still they felt no terror; rather pleasure" (chapter 51, "Spirit Spout). But once they are in action, the harpooners respected the power and the majesty of the giant sperm whale. The largest whales can stand comparison with the biggest dinosaurs, and whales are still with us—they have been with us for thousands of years.

Etymology and Extracts

Let us for the present leave the consumptive usher and the sub-sub-librarian at their tasks and continue with the "Etymology" and "Extracts" that follows these prefatory remarks. In the "Etymology," together with the several different terms for "whale," I turn to *Webster's Dictionary*: "This animal is named from roundness or rolling; for in Dan. *Hvalt* is arched or vaulted." The picture given is that of a right whale, which is considerably less of a creature than the sperm whale, hardly worth the effort to chase and kill: "All hands commonly disdain the capture of those inferior creatures," I write in chapter 73, "Stubb & Flask Kill a Right Whale." The right whale is caught mainly to balance the Pequod, which had a sperm whale tied on the other side. In my chapters 74 and 75, I continue with the comparison. After describing the elegance and power of the sperm whale's head in chapter 74, in 75, I do not hold back my disdain for these lesser beings: "The noble Sperm Whale's head may be compared to a Roman war-chariot … the Right Whale's head … to a gigantic galliot-toed shoe." I thus anticipate this difference between the sperm and right whale in the "Extracts," always recalling both the great superiority of the sperm whale, and the bravery of those who hunt them: "Myself have agreed to try whether I can master and kill this Spermaceti whale, for I could never hear of any of that sort that was killed by any man, such as his fierceness and swiftness."

By the 1800s, our American whalers were known (for the most part) to be the only ones brave enough to pursue sperm whales: "It is generally well known that out of the crews of Whaling vessels (American) few ever return in the ships on board of which they departed." I thus recount the famous and true sinking of the whaling ship, Essex, by a sperm whale, and there are similar incidents recounted, "Being once pursued by a whale which he had wounded ... the furious monster at length rushed on the boat; himself and his comrades only being preserved by leaping into the water." Furthermore, I let my reader in on the importance of the town Nantucket before it appears in the book proper: "Nantucket itself," said Mr. Webster, "is a very striking and peculiar portion of the National interest." Then too, the distinctive feature of the Nantucket whaling boat, the "specksynder," which is mentioned throughout and given its own chapter, 33, "The Specksynder" is also anticipated in my "Extracts": "It is impossible to meet a whale-ship on the ocean without being struck by her appearance."

Indeed, dear reader, even the "whiteness" of the whale, Moby Dick, is not unique: "One of our harpooners told me that he caught a whale that was white all over." To repeat, the main purpose of the "Extracts" is to warn the reader that hunting sperm whales is dangerous; these whales are huge, some with unique characteristics, and, in the encounters with them, ships are sunk and men are lost. *Given all that has already been recorded between ships and sperm whales, the reader must thus seek elsewhere for the distinctiveness of the adventure to be unfolded.*

Grammar School Ushers

I ask you, dear reader, to now reflect on the parenthetical remarks that introduce these two sections, because, in a sense, I made them stranger than what they introduce. Yes, I have written a big book, but I want you to stay awake. The "Etymology" is introduced with the parenthetical subtitle "Supplied by a late Consumptive Usher to a Grammar School." You may ask, "Does a Grammar School have Ushers?" Indeed, they are assistant teachers, and very underpaid. Well, of course, the poor men are consumptive. (Most of our universities thrive on underpaid, fully qualified teachers, whose consumptive existence supports the full professors.) But then, most writers are consumptive ushers, leading themselves to and from their writing desks. How many have had to be real ushers in order to support their writing addiction? And, if we turn our attention to the books they may have published, how many have been overlooked during the author's lifetime. I thus wonder about this long book of mine, and in chapter 32, "Cetelogy," I write, "This book is but a draught—nay, but the draught of a draught. Oh, Time, Strength, Cash, and Patience!" Yes, time, strength, and patience I can draw from within—but cash?

We Are the Temporary Custodians of Books

> But gulp down your tears and hie aloft to the royal mast with your hearts; for your friends who have gone before are clearing out the seven-storied heavens. (Extracts)

I do not know the fate of my new book, the many pages with marks and spaces on them, now by this printer and then, who knows, later by another, and, perhaps, then with pictures. Who can tell? Still, I can imagine myself in the future dusting my book, as I hold it in my hands. We are, all of us readers and writers, the temporary custodians of books. Authors die but their books remain. A well-produced book survives almost as if it were immortal, particularly, if it is printed on acid-free paper or, as with the first printed books, on sheets of "paper" made of linen. Our consumptive usher removes the dust from the outside of his books, knowing that the meanings remain intact on the inside, inviting those gifted with literacy. And yet, it is not only the permanence of the written word that fascinates me but the mystery of their content. The wonder of writing about whales is one with the wonder of whales themselves. That my book, *Moby-Dick; or, The* Whale can exist in England, in America, and throughout the world, and that its pages can contain this very narrative about whales is as great a wonder as the whales themselves. I repeat myself? And yet, I do not seek to give a philosophical track on the nature of language; rather, the deeper my suggestion, the more humorous the exposition, or so I hope. "So fare thee well, poor devil of a Sub-Sub, whose commentator I am." The sub-sub is dusting the books of Dante, Shakespeare, and Cervantes, as well as those of Plato, Aristotle, and Kant. Will my *Moby-Dick* be there also?

Meaningful Marks

What *is* that forging of marks into meanings that delivers to us a narrative? Is this question too serious for our sub-sub (or his commentator)? If Gabriel is to move aside for him, perhaps the question has not passed his notice. Letters, spaces, and punctuations, proper marks in proper order, deliver meanings! The marvel of literacy is a gift delivered to us by countless human efforts performed over several thousand years! In chapter 10, "A Bosom Friend," I have Ishmael enter the room to find Queequeg "wholly occupied with counting the pages of the marvelous book (about whaling)." Ishmael hesitates and then sits alongside Queequeg, "to explain to him the purpose of the printing, and the meaning of the few pictures that were in it." This reference to writing continues throughout my narrative; for example, I mention black letter writing, which was the early form of printing. And, I give a long comparison of whales with books in chapter 32, "Cetology," which deserves some reflection here, in this early part of my discussion.

My comparison, if properly understood, takes us back to the early days of printed books, the so-called incunabula period from 1455 to 1500. The first printers used large rectangular sheets of linen paper to print upon, and, if these were folded only once, the book was called a folio. A folio is the largest of our books, and since the original sheet folded can be of different sizes, a very large folio is sometimes called an "elephant" folio. I identify the largest whales, such as the sperm and right whales, with folios.

A folio can itself be folded, and then the book is said to be printed as a "quarto," from the Latin term "quartus," for four or fourth. In a note in the text, I explain why I do not put any of the whales in this category, namely, because of the shape of a quarto. A quarto-sized book is generally square, while a folio is a rectangle, and so, in general, are the other sizes of a book. When compared to the larger whales, the smaller whales retain "a proportionate likeness to them in figure," while this is not true of the quarto volume. Thus I go immediately to the "Octavo," which is the folio sheet now folded eight times; this is the normal size of our average printed book—if generously printed, with a nice cover—about six inches wide and nine inches long. I identify the whales of "middling magnitude" such as the black fin and killer whales with octavo books. Finally, we have rather small books, the duodecimo, three inches by five, with small print. In this class, I put the porpoises.

Am I making fun of our various ways of classifying things? Perhaps. A narrative may have many levels of meaning, but the literal should not be bypassed, there is enough mystery in that literal meaning. For what is a book physically? It is a proper arrangement of marks and spaces. Consider the letters w-h-a-l-e and their joining in the word meaning "whale." My *Moby-Dick; or, The Whale* is a remarkable arrangement of alphabetic marks, spaces, and numerous, very numerous, punctuation marks. Resting on shelves or being held by hands throughout the world, is the book, *Moby-Dick*, less a wonder than the great whales sounding the oceans of the world?

CHAPTER 4

A Tale Twice Told

Abstract If you glance over the Table of Contents listing the 135 chapters of *Moby Dick* you might be struck by the commonness of most of the titles: "Carpet Bag," "The Chapel," "Chowder," "Sunset," "Dusk," "The Whale as a Dish," "The Tail," "The Carpenter," The Hat," "The Chase—Third Day." There are, however, exceptions, and the most striking is the title of the first chapter, "Loomings." It is a mysterious term to give to the first chapter of a book. If "Counterpane" is merely a somewhat obsolete word for quilt, "looming" is defined by *Webster's* Third International Dictionary as "a mirage in which images of objects below the horizon appear in distorted form." This is fairly close to the nautical sense in which distant objects are seen reflected in the water in certain conditions. Granted, an author does not have to follow the dictionary meaning of a term, one would think that "Going Whaling" would be a more appropriate title to convey the meaning of the chapter.

But, then, yes, "Loomings," does fit. This is a tale twice told, and much of it may be dimmed by the double narrative. Melville, of course, is the author; but he writes of a story already written by one of his characters, Ishmael. In chapter 102, we learn that Ishmael had many other voyages before he took time out to write the tale we have. In one of these voyages, he visited a friend, King Tranquo, who had a temple made of the skeleton of a giant sperm whale, and lest he forget its dimensions, Ishmael had them tattooed on his right arm, leaving the rest of his body "a blank page for a poem."

J. S. Catalano, *In Pursuit of Moby-Dick*,
https://doi.org/10.1007/978-3-031-40357-6_4

"Call me Ishmael." Are not these opening three words as strange as the haunting title, Loomings?" The name "Ishmael" has Biblical references, but that is not what I see. I see three words a universe apart from the simple claim, "My name is Ishmael." Is this person really named "Ishmael," or, are we to assume that he adopts this name for the tale he is about to tell us?

Keywords Loomings • King Tranquo • Ishmael • Naming

If you glance over the Table of Contents listing the 135 chapters of my *Moby*-Dick, you will be struck by the simplicity of most of the titles: "Carpet Bag," "The Chapel," "Chowder," "Sunset," "Dusk," "The Whale as a Dish," "The Tail," "The Carpenter," The Hat," "The Chase—Third Day." There are, however, exceptions; and the most striking is the title of the first chapter, "Loomings." I grant that it is a mysterious term to give to the first chapter of a book. If "Counterpane" is merely a somewhat obsolete word for quilt, "looming," in its nautical sense, refers to distant objects reflected in the water that are made to seem real. You might think that "Going Whaling" would fit more comfortably with my other chapter headings, with their obvious sign-post character.

Why did I begin an adventure tale about whales, poetry, and *truth* by referring not to things but to the reflections of things? It is important to realize that my adventure is a mixture of the factual and the metaphysical, of what has happened and what might have happened, and, indeed, of the temporal break within the narrative that is to continue throughout the narrative. Thus, in this first chapter, Ishmael simply wants to go whaling—he is "tormented with an everlasting itch for things remote"; and, yet, this same Ishmael is also taken up with reflecting upon the way fate decides one to be born in a humble situation or amidst power and wealth—"doubtless, my going on this whaling voyage, formed part of the grand programme of Providence."

Chapter One: Loomings

The intimidation suggested by the title, "Loomings" is a proper introduction to this narrative, in which necessity and freedom are intertwined throughout, in all my numerous reflections. My tale is, in fact, constantly overlaid with reflections that could only arise from a seasoned and sophisticated sailor who has had many whaling voyages. On one level, the reason is fairly obvious, namely, I wanted my main character, Ishmael, to have the kind of knowledge about whales and whaling that would be impossible for him on his first voyage. Ishmael, thus, has two voices in my text, one a straight narrative voice that usually proceeds chronologically, and the other, a voice that intrudes itself within the tale, warning us of what is to happen—because Ishmael knows what is to happen! *The tale is over before he writes about it!*

These two voices usually fit comfortable together; but, at other times, I draw my reader into a deeper reflection. I have in mind those chapters that emphasize prophesies, hints of evil—portents. They first appear in chapter 16, with Queequeg's idol, Yojo, then in 19, "The Prophet," with Elijah; they are part of the fiber of the first mate, Starbuck, and then, later, they reappear with the strange detailed prophesies of Ahab's own harpooner, Fedallah.

Yes, I know that whalers are generally superstitious; still, for the most part, I have Ishmael handle this characteristic as part of the narrative. At other times, however, the superstitious element becomes excessive; and I want my reader to continually ponder the dark innuendo of the term "Loomings." At first, the reflection here concerns the mystery of water and the tale of Narcissus who became so enticed by his own image in the water that he plunged after it. On the other hand, the book is filled with such reflections, and there are 134 more chapters to follow!

Call Me Ishmael

And so, with the title, "Loomings," I begin this first chapter with the words,

"Call me Ishmael." Are not these opening three words as strange as the haunting title of this first chapter, "Loomings"? The name "Ishmael" has Biblical references, but that is not what I mainly intend. I see three words to be a universe apart from the simple claim, "My name is Ishmael." Is this person really named "Ishmael," or, are we to assume that he adopts this name for the tale he is about to tell us? No claim is here offered; rather, we are handed a personal engraved invitation, with perhaps a warning not to be too curious. And also, we are invited to reflect: "What after all is a name?" "Is Ishmael Ishmael?" "Is Ahab Ahab?" Yes, but also no. Who knows what change the future may bring to our lives? The narrative does not encourage us to be stuck in such reflections, and it flows into everyday prose, with a dash of humor: “Some years ago—never mind how long precisely....”

I want my Ishmael to be a strange sort of a fellow. He tells us in all seriousness that, when he is so depressed that he might actually knock someone’s hat off, he takes to the ocean, rather than commit suicide. Well, Ishmael, I think that such light-hearted depression could be dealt with by a brisk walk around the block or a glass of wine, rather than a three-year whaling voyage in which one risked one’s life. A safe and comfortable ocean voyage is another matter, but that requires money.

On the other hand, the Earth is a world of oceans with floating lands, and all life seems to have come from water. Perhaps, then, my Ishmael rightly praises water: “We see ourselves in all rivers and oceans. It is the image of the ungraspable phantom of life; and this is the key to it all.” And then too, whaling is a unique experience, even for a seasoned sailor.

Bulkington

“Call Me Ishmael.” I offer friendliness and yet distance, while I prepare my readers for the birth of a tale. Why does my Ishmael take to the sea? “Chief among these motives was the overwhelming idea of the whale himself. Such a portentous and mysterious monster roused all my curiosity. Then the wild and distant seas where he rolled his island bulk ... I love to sail forbidden seas.” Now I am letting my readers know my Ishmael. Ishmael cannot for long stand land. I will have him write a tale in which his whale ship, Pequod, is destroyed by the giant white sperm whale, Moby Dick; and that he alone survives to write about the tale. One would think that he would have had enough of whaling and return to land to write about the event in safety and comfort. No, he goes again and again on dangerous voyages before he finally decides to take time to write about his first whaling adventure. *My careful reader will note all this is clearly in the text.* This astute reader will recognize that Ishmael writes of himself, when in chapter 23, “The Lee Shore,” I have him describe the sailor named Bulkington. I have mentioned this earlier, but allow me to repeat, so that my long tale is all before you, with unnecessary strain on your memory and also because I like to write about what I think I have done well. Ishmael had seen Bulkington earlier in the Spouter-Inn (chapter 3) and noticed that he stood aloof from his fellow sailors, although, indeed, he was very much liked by them. They had just returned from a long voyage, and the Spouter-Inn was their first stop. Ishmael had not yet picked out the Pequod as his whaling ship, but that voyage was only a few days away. Now, on board, he sees Bulkington, and, in this very short chapter of only a page and a half (chapter 23), I have him observe, "I looked with sympathetic awe and fearfulness upon the man, who in midwinter just landed from a four years' dangerous voyage, could so unrestingly push off again for still another tempestuous term."

Bulkington does not appear again, but his presence helps us to understand Ishmael and, retrospectively, the first chapter, "Loomings." The notion of the "lee shore" is filled with poetic and philosophical insight. Normally, a ship at sea looks to the shore as a haven; but, in a storm, that same shore can be the greatest danger—better to go out further into the sea than be dashed upon rocks on land. And so for the true sailor, land is itself a danger, too much comfort, too little space, too uneventful for true thought: "Know ye, now, Bulkington? Glimpses do ye seem to see of that mortally intolerable truth; that all deep, earnest thinking is but the intrepid effort of the soul to keep the open independence of her sea...."

Yes, dear reader, every ship is a world unto itself. Life therein is more easily and carefully defined than on land. Even in a pleasure cruise, meals, recreation, and status itself is fixed. In a merchant ship, war ship, or, as here, a whaling ship, one also finds the meaning of life reflected in the ship itself, temporarily keeping at bay some of the difficult questions of human existence: Does God exist? Perhaps, yes, perhaps no; but the captain exists and on this voyage he is a god, aloof for the most part, while the mates, harpooners direct the common sailors, each of whom knows how their part fits within the whole. Compared to the grimy, slavish jobs that many of these sailors would find on land, their roles here are indeed those of demigods! "What of it, if some old hunks of a sea-captain orders me to get a broom and sweep down the decks? ... Who aint a slave? Tell me that" (chapter 1, "Loomings"). These "slaves" have at least freely chosen their roles.

Chapter 102, "A Bower in the Arsacides"

"The great flood-gates of the wonder-world swung open ... endless processions of the whale, and midmost of them all, one grand hooded phantom, like a snow hill in the air" (Loomings). Moby Dick is that snow hill. You must forgive my Ishmael for mentioning this whale in this first chapter, because, after all, to repeat again the key to my tale, all the events have already happened to him. We must expect him to now and then anticipate the outcome of the narrative he is relating to us. On the other hand, I grant that this anticipation will seem, at times, to break all normal bounds for the sake of its twice-told nature.

And, as if it were a key to a great treasure, allow me to jump ahead to the strangest chapter in my entire book, chapter 102, “A Bower in the Arsacides.” When you reach chapter 102, you might want to look up information about Arcacides; but don’t bother, it is really an invitation to relax a little and congratulate yourself for reaching chapter 102! Ishmael is speaking, explaining to us how he knows so much about the size of sperm whales. This information he informs us was obtained after the tragic voyage of the Pequod, of which he was the only survivor. He had many other voyages before sitting down to write the tale we read. On one of these voyages, he was invited by his friend Tranquo, “king of Tranque, itself one of the Arcadies,” to spent time at his villa. (Are we to believe that this king was related to the ancient King of Parthia, 250 B.C., founder of the dynasty which, although overthrown in 226 A.D., nevertheless still held true to its older kings? Stay with the humor.)

On the Island, Ishmael sees a temple made of a large sperm whale’s skeleton, not killed but beached by a storm. Before he enters, he notices the way living trees and living earth weave about this great skeleton, and the way the light of the sun plays about it, “The great sun seemed a flying shuttle weaving the unwearied verdure.” Does most of the chapter reads like a sermon? But then much earlier my Ishmael had devoted chapters 7 and 8 (“The Chapel” and “The Pulpit”) to prepare us for chapter 9, “The Sermon,” given by Father Mapple. Both are about a whale, and indeed, Jonah is also mentioned in 102 but only in passing. But to continue, there is more strangeness to follow.

Ishmael enters the whale-temple to measure its size. The priests assert their rights, arguing over feet and inches. During their argument, Ishmael manages to slip away and copy the dimensions down. But lest he forget them, and because he wishes to share this information about the gigantic size of whales with you when he writes his book, he has all the information tattooed on his right arm, leaving the rest of his body “a blank page for a poem.” Is that poem yet to come, or does Ishmael give it to us now as he prepares us for the following chapters detailing the size of whales?

> Now, amid the green … the great white, worshipped skeleton lay lounging—a gigantic idler! Yet, as the ever-woven verdant warp and woof intermixed … Life folded Death; Death trellised Life; the grim god wived with youthful Life, and begat him curly-headed glories.

You may ask if this whale temple is any different from a wooden fence, intertwined with vines and roses. Here, and throughout my narrative, I invite my readers to become aware of the marvels of our everyday life, through a third

voice hinting that the wildness of his later adventures may have a part in the tale he is unfolding. Memory is not an innocent affair.

On Normal and Excessive Humor

Recollection of the past is always affected by the present. At best, we take this awareness into account in our attempts to remember exactly what happened earlier in our life. For the most part, I let my Ishmael do just that. He lets us know that he is attempting to recall his first whaling experience—although this event occurred much earlier in his life—and that later he has had many other voyages. Indeed, he openly informs us about some of those later voyages, while he is also recounting his first voyage aboard the Pequod.

We thus have a "twice-told tale." It is told first, in the ordinary sense of a narrator telling us a tale that occurred earlier in life, while also informing us of later related adventures. That is, while the narration mostly occurs in the present tense, the entire context is given as a recollection, within which past and present are mixed. For the most part, our Ishmael does this throughout the narrative in such a way that we know where we are, and it works well. But then, I want my reader to wonder whether there is another, deeper sense in which this narrative is twice-told.

Let me first draw attention to an "excessive" humor that appears now and then in my tale. You will see that there is a constant humor throughout the narrative—despite Ahab's monomania. Usually, I have this humor to lighten the reflections when they tend to become too metaphysical or the tale itself overly serious too soon. For example, consider the light chapter 101, "The Decanter," which occurs right before my strange chapter 102, "A Bower in the Arsacides," examined directly above. Chapter 101 concludes, "If you can get nothing better out of the world, get a good dinner out of it, at least. And this empties the decanter." In a sense that would be a good picture of my whole book, as my friend Philip Bishop has shown, and he thinks that it should be put on the cover of a future edition.

At other times, however, I grant that the humor may be so excessive that it might seem to put in question the truth of the narrative being related to us. Aside from the strange humor in chapter 102, "A Bower in the Arsacides," I invite you to consider my chapter 85, "The Fountain":

> And I am convinced that from the heads of all ponderous profound beings, such as Plato, Pyrrho, the Devil, Jupiter, Dante, and so on, there always goes up a certain semi-visible steam, while in the act of thinking deep thoughts. While composing a little treatise on Eternity, I had the curiosity to place a mirror before me; and ere long saw reflected there, a curious worming and undulation in the atmosphere over my head.

Yes, the humor somewhat fits the chapter about reasoning whether the spout of the whale is water or mere vapor. But, in fact, I have it move beyond that need. True, we have reached chapter 85, and there has been a great deal of heavy reflections, and a dash of humor is due the faithful reader. But this is more than a "dash." And, again, this occurs in the very form in which chapter 54, "The Town-Ho's Story" is related. True, Ishmael states, "For my humor's sake, I shall preserve the style in which I first narrated it at Lima." But this admission will not, I hope, excuse the extent of the humor. As we proceed below, I will be pointing out other instances where it should seem to my reader that my Ishmael is altering the tale he is relating; or, at least, giving us doubts that parts of it may be exaggerated.

We are thus back to the title of the first chapter, "Loomings," for what looms ahead dimly, both for Ishmael and my readers, is the narrative itself, captured and fixed by the written word as a camera may capture a mirage. Mirages are not mental fictions; they can be photographed. They exist for all to see, just as the written word exists. If there is "error" in a mirage it is one of judgment, namely, to misjudge the image of a ship for a real ship. A reflected image of you exists in a mirror; the image is not false as an image; it is only false if it is mistaken to be the real you, standing in front of the mirror.

And so, I am giving my readers two tales, one, the real adventure about whaling—a unique adventure within which Jonah, Plato, and Shakespeare all seem to be riding on the backs of those great leviathans, whales; and another adventure, its reflection, somewhat altered, which both is and is not the truth of the first. Thus, you must be prepared for the strangeness of the conclusion of this tale. We will see that Ahab dies *exactly* as Fedallah predicts, in all the three aspects of his prophesy. And Ishmael survives on a coffin made for his friend, Queequeg, who regained his health and did not have to use it. All die except Ishmael, who is to be our narrator. As a questioning reader, this should not sit well with you. True, I have written what is called fiction. Fiction, indeed, but this fiction aims to give us the *truth of whales and whaling—and, for the most part, it does give us that truth.* What this other "twice-told" aspect of the tale—this reflected aspect—might mean for the reader, what the *truth of my written conclusion might mean, is a wonder not easily resolved, and not meant to be easily resolved.* Yes, I will offer a resolution; and, yes, you are invited to have your own.

CHAPTER 5

Ishmael and Queequeg

Abstract Ishmael finally seeks cheap but comfortable-looking lodgings in "The Spouter-Inn." He is told that he must share a bed. His companion turns out to be Queequeg, the harpooner, who is to remain with him until the final chase after Moby Dick and whose "coffin" is to save Ishmael's life. In chapter 12, "Biography," we are told that Queequeg is a native of "Kokovosko," an Island far away to the West and South. It is not down in any map.

I wish to return briefly to the earlier chapters, 7, 8, and 9: "The Chapel," "The Pulpit," and "The Sermon." Here we are given reflections on life and death. Again, I question the bleak critical outlook on *Moby Dick*. Melville has control of his tale. Why does he devote three chapters to a church and the sermon given? "To give the tale atmosphere," one might say, and to point to the dangerous lives of whalers. Yes, but does that require these three chapters, with its appealing picture of Father Mapple and his sermon. Then too, Queequeg is frequently praying to his little black god, Yojo. Yojo's recommendation is to be crucial to the tale.

Keywords Queequeg • Pulpit • Father Maple

(Chapters 2–15 of my Moby-Dick)

J. S. Catalano, *In Pursuit of Moby-Dick*,
https://doi.org/10.1007/978-3-031-40357-6_5

From the mirror and mist suggested by the title of the first chapter, "Loomings," I now have my readers begin our journey in search of whales on a clearly lit path—at least as most of the chapter headings are concerned. In reading my complex book, filled with literary, philosophical, and religious reflections, it will be useful for my reader to keep reminding oneself of the down-to-earth character of most of the chapter headings. If there are a few that are mysterious, most have an extraordinary commonness:

Chapter 2, "The Carpet-Bag"
Chapter 3, "The Spouter-Inn"
Chapter 4, "The Counterpane" (Quilt)
Chapter 5, "Breakfast"
Chapter 6, "The Street"
Chapter 7, "The Chapel"
Chapter 8, "The Pulpit"
Chapter 9, "The Sermon"
Chapter 10, "A Bosom Friend"
Chapter 11, "Nightgown"
Chapter 12, "Biographical"
Chapter 13, "Wheelbarrow"
Chapter 14, "Nantucket"
Chapter 15, "Chowder"

Further, the everyday chapter headings are my reminder that what is true for whalers is also true for each and every person on the planet. The reader should not consider this interpretation, for the relation of the dangerous life sailors pursuing sperm whales, while in its own way unique, has its counterpart in the most ordinary life lived on land. Events do not have to be extraordinary to be filled with significance.

> However a good laugh is a mighty good thing, and rather too scarce a good thing … And the man that has anything bountifully laughable about him, be sure there is more to that man than you perhaps think for. (Chapter 5, "Breakfast")

On the other hand, although the chapter headings are remarkable evident, some mist from that first chapter, "Loomings," haunts the contents, not only these early chapters but of the tale as a whole. For example, on seeing the sign of a friendly looking inn in chapter 3, "The Spouter-Inn," Ishmael thinks, "The Spouter-Inn—Peter Coffin. Coffin?—Spouter?—Rather ominous in that particular connection, thought I." And, as we will see, Queequeg's body is strangely tattooed all over; much later in the tale, he is willing to talk about it. In chapter 110, "Queequeg and His Coffin," I have Queequeg, after having a coffin made for his dying body, to recover because he remembers that he had an obligation to fulfill. In the leisure of his sickness, his tattooed body becomes all the more evident to the sailors. Queequeg mentions that they were done by "a prophet and seer of his island." They were made to explain the relation of man to the universe, "a mystical treatise on the art of attaining truth." I then have my Ishmael, noticing Ahab looking at Queequeg's body, reflecting that they must have been the reason for Ahab's words, "Oh, devilish tantalizations of the gods!"

But every human body contains such a treatise; although, except in sickness, we are too accustomed to its marvel. The human body is central to my tale, as central as that of the great sperm whale itself. Throughout my *Moby-Dick*, the movements of the human body while at work are described as carefully as the massive bodies of the sperm whale swimming the oceans of the world. Each sailor has his tasks, and the movements of those tasks are one with the rhythm of the ship itself. In my chapter, 72, "The Monkey-rope," I observe, "There is no staying in any one place … It is much the same with him who endeavors the description of the scene." This reflexive awareness of writing, within the narrative being written, begins in chapter 2, "The Spouter-Inn." My Ishmael is reflecting upon an old text, "True enough, thought I, as this passage occurred to my mind—old black letter, thou reasonest well." "Black-letter" printing is characteristic of the early printed book, certainly the incunabula, the first printed books from 1455 to 1500, but it may also extend to the early 1600s. This passing reference in chapter 2 would not justify my expanding upon it, except that references to the very format of the book recur throughout my long tale. I have already mentioned the long comparison of whales to the format of books, in chapter 32, "Cetology," a comparison which takes us back again to the very origin of making books. And, to repeat, it continues throughout the tale; for example, chapter 45, "The Affidavit," begins: "So far as there might be of narrative in this book." In *Moby*-Dick, the majesty of whales and the miracle of writing about them go hand in hand.

Meeting Queequeg

'Landlord, for God's sake, Peter Coffin!' shouted I. 'Landlord! Watch! Coffin! Angels! save me!' (chapter 3, "The Spouter-Inn")

My Ishmael has decided to enter the Spouter-Inn, despite the ominous name of its owner, Peter Coffin. It seems friendly and cheap, and it is crowded. He is told that he must share a bed with a harpooner who has not yet arrived. Ishmael is not pleased. He remarks that although sailors share the same sleeping quarters, they each have their own bunk. But, it is cold and damp outside, and he accepts.

Tired of waiting for the harpooner to arrive, Ishmael finally goes to bed. The harpooner finally arrives and begins to undress. Ishmael is spellbound by his appearance and his unique preparations for bed. He hesitates too long to announce his own presence, and the cannibal-harpooner takes to bed, smoking his tomahawk pipe. Sensing Ishmael's presence, however, he begins to threaten Ishmael with his tomahawk. Finally, Ishmael calls out to the landlord, who arrives smiling, calming Ishmael and informing the harpooner, named, Queequeg, that he must share his bed. All ends, with Ishmael reflecting in the same chapter, "I turned in, and never slept better in my life."

Ishmael's friendliness with Queequeg is a mixture of openness and mystery. Besides the mystery of his tattoos, there is Queequeg's close bond with his black idol, Yojo, as well as his strange ability to recover from near death. To anticipate again my chapter 110, "Queequeg in His Coffin," Queequeg has caught a fever and seems on the verge of dying. He does not wish to be buried in his hammock as was the custom at sea, but desires a true water-tight coffin to be made. The carpenter agrees and finishes his work. Queequeg tries the completed coffin for size and seems ready to die. But, at the last minute, Queequeg remembers that he has left a duty on shore undone, and he recovers. Ishmael reflects, "It was Queequeg's conceit, that if a man made up his mind to live, mere sickness could not kill him, nothing but a whale or a gale, or some violent, ungovernable, unintelligent destroyer of that sort."

Ishmael's coffin continues to play a crucial role in the narrative. In chapter, 126, "The Life-Buoy," upon the loss of the ship's own life-buoy, Ishmael's coffin serves that purpose: "A life-buoy of a coffin! cried Starbuck," the first mate. "'It will make a good enough one,' said Flask," the third mate. And, of course, it is to be that same coffin-life-buoy which will save Ishmael, when all others are lost in the sinking of the Pequod by the giant white sperm whale, Moby Dick. Warnings, portents, mysteries abound in my tale from its beginning to its end. I want my reader to wonder about it all; and yes, about the ending I have written for my owners who are supporting my writing and to that other ending I hint at to satisfy my conscience.

Strange Sensations

Ishmael has fallen asleep with Queequeg and awakens to find a tattooed arm over him, its pattern melting into the patchwork of the quilt. He recalls an incident in childhood, when, after misbehaving, he had to go to bed in the bright afternoon during the longest day of the year, June 21. He begged to be excused and whipped instead, but to no avail. Finally, he falls asleep and awakens at night: "Instantly I felt a shock running through all my frame; nothing was to be seen and nothing was to be heard; but a supernatural hand seemed placed in mine." We might think that his mother had returned to comfort him, but Ishmael thinks the event is filled with more mystery: "Nay, to this very hour, I often puzzle myself with it."

But here, dear reader, do not pass over a commonsense explanation for this frequent reference to omens. Ishmael is writing long after the events aboard the Pequod have happened. *After* an unfortunate happening, it is frequently possible to see warnings that we should have headed. And, *after* a successful adventure, it is also possible to think back on warnings that we fortunately did not heed. Superstition can be part of life; and, perhaps, it is particularly so for sailors on a long voyage. But, perhaps while true, this explanation is too general to satisfy you?

Closer to the tale itself is the drama between free will and the circumstances of life; circumstances, which, at times, seem to be laden with a mysterious "fate." Why is one person born blind or deaf and another healthy; and, why is one person born within a poor or rich family, with loving or unloving parents? Congenital blindness and deafness seem, at the present time, to be beyond our control; but, the poverty and the lack of love that many people experience may reflect our individual and collective selfishness and evil.

Am I, perhaps, being too clear for the intent of my novel? Reflecting back again to the first chapter, "Loomings," we are explicitly informed of the strangeness of providence: "I cannot tell why it was exactly that those stage managers, the Fates, put me down for this shabby part of a whaling voyage." But, as soon as Ishmael begins to weave for us a tale of fate, he then provides for us the very reasons for actions, "Chief among these motives was the overwhelming idea of the great whale itself." This tension between freedom and necessity, choice and circumstance continues throughout my narrative.

As the tale continues on board the Pequod, the circumstances that define the limits of one's freedom are relatively fixed, even if the reflections that occur on the journey are not. Disasters and mutinies are all possible on the voyage, and the drama among peoples is here relatively fixed, namely, between the mates, harpooners, and Ahab, on the one hand, and the crew, on the other. But, as always in life, there is the planet, Earth—the vast oceans and the whales therein; and there is the weather.

In chapter 29, "Enter Ahab; to Him, Stubb," in which not only the title but the dialogue and soliloquy seem to reflect Shakespeare, the very brief chapter begins with the power, magic, and mystery of air, sky, and water to affect the mind: "But all the witcheries of that unwaning weather did not merely lend new spells and potencies to the outward world. Inward they turned upon the soul." And thus, the weather seems to adversely affect Ahab. He has been walking the deck, his whale-bone leg pounding the ship's planks. At times, he seems aware of the racket and sleepless nights his walking brings to the crew, and other times he does not seem to care. Stubb, the second mate, thinks he might casually remind the captain that the sailors need their rest. But there is nothing casual about the encounter: "'Am I a cannon-ball, Stubb,' said Ahab .... 'Down, dog, and kennel!' 'I will not tamely be called a dog, sir.'" But Stubb ultimately backs down. The seed for possible mutiny has been sowed, but it never bears fruit.

On the other hand, the weather almost tames Ahab, bending him away from his evil fixation on Moby Dick: "Oh, Starbuck! it is a mild, mild wind, and a mild looking sky. On such a day—very much such a sweetness as this—I struck my first whale—a boy-harpooner of eighteen!" (chapter132, "The Sympathy").

Freedom and fate, understanding and bad faith, in brief, the individual and the world with all that is within it contribute to weave a web in which clarity and omens exist side by side throughout my *Moby-Dick,* until the very end of the tale. Thus, in chapter 117, "The Whale Watch," Fedallah, Ahab's mysterious harpooner, also called the "Parsee," informs Ahab that he will not die until he sees Fedallah go before him: "I shall go before thee thy pilot." And further:

> But I said, old man, that ere thou could die on this voyage, two hearses must verily be seen by thee on the sea; the first not made of mortal hands; and the visible wood of the last one must be grown in America.

In chapter 134, thinking about these predictions, Ahab says to himself, “These things called omens! … *I’ll, I’ll* solve it, though!” On the contrary, Ahab watches as every aspect of Fedallah’s predictions are coming true before his eyes; and he nevertheless, continues in his mad pursuit after Moby Dick. Thus, Fedallah is indeed seized by Moby Dick as he is trying to harpoon the whale. The rope from the harpoon binds him to Moby Dick; and Fedallah rises from the water with his tattered remains tied to Moby Dick—the hearse not made by mortal hands. Then, I have Moby Dick sink the Pequod, and Ahab realizes that its wood can only be from America. Finally, my Ahab here dies by the hemp of the rope from his harpoon as predicted.

What are you to make of all these prophecies? Are they indeed part of the more or less chronological tale that I have my Ishmael honestly attempting to recall, with harmless references to his past? Or, are they, dear reader, insertions, more than innocent additions to the tale being narrated, and are we again back to a deeper sense of the twice-told nature of this tale? For the present, however, we must move on with the adventure.

The Whaleman’s Chapel

> In this same New Bedford there stands a Whaleman’s Chapel, and few are the moody fisherman, shortly bound for the Indian Ocean or Pacific, who fail to make a Sunday visit to the spot. I am sure that I did not. (chapter 7, “The Chapel”)

I give you not one, but three chapters devoted to this whalesman’s church, the chapel itself, its pulpit, and the sermon given therein, namely, Chapters, 7, 8, and 9, all with its obvious titles, “The Chapel,” “The Pulpit,” and “The Sermon.” I have already mentioned the possible relation between these three chapters and the chapel made from the skeleton of a sperm whale in chapter 102, “A Bower in the Arsacides.” But the fact remains that I do devote three chapters to a rather straight-forward religious architecture and sermon. True, my reflections within the chapters are not ordinary, but typical of my poetic genius, if I do say so. These three chapters are both convenient steps as I build my tale of blubber, poetry, and truth; but there is more.

The architecture of the church and the sermon given within it, reflect the architecture of the Pequod and the adventures of whalers lived during its voyage. Before elaborating this analogy, let us first consider, dear reader, how the biblical story of Jonah delivered as part of the sermon by Father Mapple fits our tale.

Jonah's "great fish" has always been interpreted as a whale, and in chapter, 83, "Jonah Historically Regarded," we have an entire (short) chapter devoted to a theme that is mentioned several times in my *Moby-Dick*. Ishmael considers the pros and cons of believing in the tale of Jonah, and his final attitude seems to be that, true or false, the tale of Jonah exists and it is part of our heritage.

Let us consider the sermon itself, given in this relatively long chapter, 9, "The Sermon." Father Mapple recounts the short book, "Jonah," from the Old Testament, "Shipmates, this book, containing only four chapters—four yarns—is one of the smallest in the mighty cable of Scriptures." Father Mapple's summary is seemingly straight-forward, although he concentrates on the relations between the crew and Jonah. Mapple thus skips the details of Jonah's disobedience, namely, that God had ordered Jonah to admonish the sinful people of Nineveh and call them to repentance. Jonah, however, attempts to avoid the task, and he boards a ship to escape God judgment. (Father Mapple, however, does call attention to the simple fact that Jonah paid his fare: "In this world, shipmates, sin that pays its way can travel freely.") A great storm arises, the ship is about to sink, and the sailors guess that one of them is a sinner. They take lots to see who is to blame. The lot falls on Jonah; he asks to be thrown overboard, and when it is done, the sea calms. A great fish saves Jonah, who repents, and then fulfills his task. Nineveh also repents and is saved. It is interesting that in the biblical tale Jonah is rather peeved that he had to go through all the trouble, only to have Yahweh forgive the people and the animals.

In the Old Testament no blame is put upon the sailors who tossed Jonah overboard. If real mutiny never occurs on the Pequod, the possibility haunts the tale until the very end: "'Oh! Ahab,' cried Starbuck, 'not too late is it, even now, the third day, to desist. See! Moby Dick seeks thee not. It is thou, thou, that madly seekest him!'" ("The Chase—Third Day").

My Only Pictures

> The wife of a whaling captain had provided the chapel with a handsome pair of red worsted man-ropes for this ladder. (chapter 8, "The Chapel")

When reflecting upon this whaling adventure, we cannot forget money. Religion, enterprise, and profit have sent these sailors on their journey. The marble tablets in the church reflect substantial contributions. In the text, we see these tablets enclosed within boarders.

These tablets are the only picture-like things in my book, with the exception of Queequeg's mark, which he uses to affix to his contract with the owners of the Pequod (chapter 18, "His Mark"). In the text, the mark is described as a strange round figure, which could be interpreted as a sketch of a whale. But the texts of the British and American first editions give a cross, perhaps because it was easier to draw, or, perhaps, because the figure in the manuscript, written by my sister, Augusta, was not clear.

You might have wanted some pictures of whales in my book; but my pictures are my descriptive words. Yes, here, I do provide pictures of these marble tablets. They are significant, pointing to the watery graves not marked by any tombstone. Thus, in the same chapter, 7, "The Chapel," Ishmael reflects: "Oh! Ye whose dead lie buried beneath the green grass; who standing among flowers can say—here, *here* lies my beloved; ye know not the desolation that broods in bosoms like these."

Do these drawn tablets aim then to mark the many unmarked dead, not only at sea but throughout the world? If so, I hope they work. These sketches on the pages of my book are less perishable and more omnipresent than those of stone that mark earthly graves. (And, then too, Queequeg's mark may stand for the scratches of all the illiterate throughout the world.)

My Ishmael's Reflections on Death

But, in truth, my Ishmael's reflection of death is not consistent. If you had to analyze them, I do not think I could come up with a clear understanding of Ishmael's beliefs. He reflects toward the end of chapter 7: "Methinks that what they call my shadow here on earth is my true substance." Fine, this is relatively clear; it seems to be in opposition to the tradition of Plato (about 429–347 B.C.), and more within the tradition of his famous pupil,

Aristotle (384–322 B.C.), who indeed put more emphasis on the body (our shadow) than Plato's emphasis on the soul (our true self).

I am not imposing philosophical knowledge into what I have written, for my Ishmael clearly informs us about his knowledge of Plato. Thus, in chapter 35, "The Mast-Head," Ishmael, to repeat, humorously warns the captains of whaling boats, "And let me in this place movingly admonish you … Beware of enlisting … any lad of lean brow and hollow eye … this sunken-eyed Platonist will tow you ten wakes round the world, and not make you one pint the sperm richer."

And, to a great extent, this reflection on Platonism seems to fit Ishmael himself; he is the one who dreams while he should be looking for whale. And, yet, in the same paragraph, he reflects: "Methinks that what they call my shadow here in earth is my true substance," and "take my body who will, take it I say, it is not me." Ishmael, which is it? Does your body hold your personality, or something else, perhaps, your spirit? Well, I have not written a philosophical tract, and our everyday reflections on life and death are not always consistent; and, indeed, the entire paragraph is philosophy infused with the poetry of water and the life within.

But, if Ishmael's thoughts are not consistent, the church, in its architecture, is indeed so, for it is itself a whale-ship on earth. In chapter 8, "The Pulpit," there is a perfectly constructed rope-ladder, which Father Mapple climbs, then lifts up, signifying his temporary "spiritual withdrawal." The chapter concludes, "Yes, the world is a ship on its passage out, and not a voyage complete; and the pulpit is its prow." A nice analogy, one that fits the theme; but then architecture is a significant factor and source of analogies throughout my book. Thus, in chapter 96, "The Try-Works," we are informed, "Besides her hoisted boats, an American whaler is outwardly distinguished by her try-works." The try-works occupy a significant area on the whaler; it is the place where the oil is gathered from sperm whale. While cleaning out one of the two deep round try-work pots within which the blubber is melted, Ishmael reflects, "I was first indirectly struck by the remarkable fact, that in geometry all bodies gliding along a cycloid, my soapstone for example, will descend from any point in precisely the same time."

Aside from the try-works, there are detailed descriptions of all the ship's physical features and equipment, with their own chapter headings. I have already mentioned "The Line," chapter 60, but there is also 62, "The Dart," and 63, "The Crotch."

Returning to chapter 96, "The Try-Works," we have rooted into the firm architecture of the ship, the dread of not knowing where one is on the planet—vast oceans upon oceans always lie before one. We now have reflections that rival the sermons of Father Mapple: "The truest of all men was the Man of Sorrows, and the truest of all books is Solomon's, and Ecclesiastes is the fine hammered steel of woe. 'All is vanity.' All." Yes, all is vanity; but Ishmael will not stay home and pray. No, Ishmael, making himself, the lone survivor of the Pequod, will, after this deadly voyage, go out on many "wild" voyages, until (too old for real adventure?) I have him land bound long enough to write this adventure.

CHAPTER 6

Going Whaling and a Hint of Ahab

Abstract Every whaling ship has a captain, but we must distinguish between Ahab as captain of the Pequod, and Ahab as he is in mad pursuit after the whale Moby Dick (Ahab as Ahab). We will see that Ahab as Ahab does not dominate this adventure. For most of the voyage, he lets the mates do their tasks. ***Indeed, if Melville wanted to write a book mainly about a mad chase after a whale, he could have retained about 45 chapters of the 135 chapters and still have plenty of text-room for reflections on life in general and life in America in the 1850s.*** Ahab's monomania does not unify *Moby*-Dick, and the effort to interpret the book in this way gives us not a classic but a mess—which is the way it was regarded by most of the early reviewers.

It was not Ahab that drew me back to reading *Moby-Dick* on and off for more than thirty years. Nor was it my philosophical career, for aesthetics was never my forte. It was the *story itself!* Poetry and truth I could find elsewhere, but not as they are one with whales. These leviathans, traveling through all the oceans of the earth as easily as we walk our pathways and streets, spout not merely vapor but a story of the planet Earth itself.

(Chapters 16–25 of my Moby-Dick)

J. S. Catalano, *In Pursuit of Moby-Dick*,
https://doi.org/10.1007/978-3-031-40357-6_6

One would think that given Queequeg's vast experience that he would go with Ishmael to pick out the ship they are to board. The chapter, however, begins with Queequeg telling Ishmael that Yojo has strongly insisted that Ishmael alone must choose the ship they are both to board. Now this ship is to be the Pequod, on which all are to die, except Ishmael, who is to be saved by a coffin made for Queequeg that he never used as such but which then became a life-raft. I will devote some space reflecting upon the significance of this for the narrative as a whole, and I will return to it as I continue. Here, I will only recall the nature of this narrative as "twice-told," namely, the events narrated being recorded by Ishmael much later than they happened, and after Ishmael had other voyages.

Keywords Ahab • Queequeg, • Yojo, • Twice Told

My dear reader, you must be prepared for both clarity and mystery in reading my hefty book. Consider how I was supposed to choose the boat we were to sail on for our long journey. Queequeg insisted that, having diligently consulted Yojo, the name of his black god, Yojo earnestly enjoined that the selection of the ship rest wholly with me. I discuss this in my chapter 16, "The Ship").

We are back to prophesies, hints of evil—portents. For the immediate present, however, we can note that these ten chapters of deciding upon a ship, boarding it, and sailing out, are *both* laced with mystery *and* marked by clarity. First, the markers of the chapter headings are mostly very clear: 16, "The Ship," 17, "The Ramadan" (Queequeg's religious fast), 18, "His Mark," 19, "The Prophet," 20, "All Astir," 21, "Going Aboard," 22, "Merry Christmas" (the day of their leaving shore), 23, "The Lee Shore," 24, "The Advocate," and 25, "Postscript." There are, of course, the usual metaphysical and poetic reflections; but these, I believe, slowly reveal themselves upon re-reading, either briefly while you are enjoying the narrative, or upon repeated readings.

What is mysterious, however, is that all these portents seem to be unnecessary, for they do not inhibit anyone's actions; and, except for Fedallah's explicit predictions about Ahab's death, they have natural explanations. For example, in chapter 16, "The Ship," it seems strange that Queequeg's god, Yojo, should insist that Ishmael alone pick out the ship on which they are to sail. At first, Ishmael doesn't like the idea; it seems clear to him that Queequeg, with all his experience, would be the better judge of a proper whaling vessel. But, when Ishmael looks at those available, there are, in fact, only three ships from which to choose: "After much prolonged sauntering and many random inquires, I learnt that there were three ships up for three-years' voyages—The Devil-dam, the Tit-bit, and the Pequod" (chapter 16, "The Ship"). Discounting the strange names of the first two, it is only the Pequod which fits Ishmael's image of how a good Nantucket whaler should appear: "Her venerable bows looked beaded."

Thus, all the mystery behind Yojo's injunction seems to have been unnecessary. Indeed, I give us such a detailed description of the Pequod that one might wonder whether either Ishmael or Queequeg would have chosen any other ship. If Yojo had said that they should board the least likely looking ship, or had Yojo hinted at the names "Devil-dam" or "Tit-bit," were safer ships to board, there might have been some purpose to the seeming mystery.

On the surface, I grant that it might seem like a badly written mystery story. Yes, you will see that I will write it so that it seems that Captain Ahab leads them to destruction. Let us also grant that my tale requires Ishmael to be on the Pequod, with Ahab as its captain. But why, you may rightly ask, do I give my readers this maneuvering, dictated by Queequeg's black idol? Of course, to repeat, superstition is part of life; it is particularly part of these sailor's lives. That is true; but I want my reason to be deeper, an aspect of the unique "twice-told" nature of the tale, which began in the first chapter, "Loomings," and which I will return to in my conclusion.

On the other hand, I have Ishmael provide us with an insight into his own leanings toward omens. In chapter 20, "All Astir," Ishmael is preparing to board the Pequod, although he has not yet seen the person who is to be his captain on this long, three-year voyage. He reflects:

> If I had been downright honest with myself, I would have seen very plainly in my heart that I did but half fancy being committed this way to so long a voyage, without once laying my eyes on the man who was to be the absolute dictator of it, so soon as the ship sailed out upon the open sea. But when a man suspects any wrong, it sometimes happens that if he be already involved in the matter, he insensibly strives to cover up his suspicions even from him-

self. And much this way it was with me. I said nothing, and tried to think nothing.[1]

Elijah 2

'Did ye see anything looking like men going towards that ship a while ago?'

Struck by this plan matter-of-fact question, I answered, saying 'Yes, I thought I did see four or five men; but it was too dim to be sure.'

'Very dim, very dim,' said Elijah. (chapter 21, "Going Aboard")

The tension between the fate and freedom, mystery and clarity, prophecy and reason becomes explicit in my chapter, 19, "The Prophet." An old sailor, who calls himself, "Elijah," approaches Ishmael and Queequeg. Elijah hints at the danger of sailing the Pequod with Ahab. Ishmael rightly says to Elijah, "If you have anything important to tell us, out with it." But Elijah merely endorses Ishmael spirit, "I like to hear a chap talk up that way."

In chapter 21, "Going Aboard," I have made Elijah's presence to be equally ambiguous. First, he asks a commonsense question about whether Ishmael has seen any men board the ship. Ishmael replies that, in fact, he thought that he saw four or five board the ship, but it was dim. Elijah then gives another hint of danger, "Oh! I was going to warn ye against—but never mind ... its all one, all in the family too."

Aside from the indirect warning about Ahab, the episode is interesting in light of what follows. In fact, Elijah was right about men secretly boarding the Pequod; they are Ahab's private crew for his boat—a private

[1]We frequently hinder and redirect the clarifying light of our own reflections upon our behavior: We choose not to know, and this choice is effective: *We do not know that which we do not which to know.* I believe that we now have a partial grasp of why I resort to omens and portents, namely, because when we cooperate in evil and refuse to face our own deeds, we begin to see the workings of fate, or of some force beyond that of our own individual powers leading us to evil and preventing us from doing good. Starbuck is inclined to believe in portents, and he is also too weak to oppose Ahab. Yet, my Ishmael explicitly hesitates to blame him. Even Queequeg seems to go along with Ahab's mad desire for revenge. And now Ishmael begins to see that he also should have been more awake to the real danger signs and less prone to give weight to omens.

Are these omens Melville's (through Ishmael) way of allowing us room to excuse the moral weakness of those on board the Pequod to revolt against Ahab? And, if so, have we solved the mystery of the persistent presence of omens within the tale of my *Moby-Dick?* No, but we have made a start.

harpooner, Fedallah, and a private crew to row his boat when chasing whales. Although Ahab was captain, the owners would never have approved of a special ship and crew, particularly, as Ahab had only one good leg. Of course, this crew cannot remain hidden for long. Although I have provided early hints that there are stowaways, they are first revealed in chapter 48, "The First Lowering," and then considered in more detail in chapter 50, "Ahab's Boat and Crew—Fedallah."

It should be clear, I think, that Elijah knew about this secret crew, and, perhaps, as a former sailor on board the Pequod, he recognized Fedallah as a harpooner with a dubious background. But can Elijah actually prevent Ishmael and Queequeg from boarding the Pequod? Ishmael and Queequeg have signed papers and they have made up their minds to board the ship. Elijah gives the best warning that he can. When Ishmael says that it was too dim to be sure whether sailors had boarded, Elijah replies, "Very dim, very dim," referring, as the tale will reveal, not merely to the light but to the men themselves.

Elijah reveals himself to be what he said he was, namely, an old sailor who had lost his arm, and who spends time observing whaling ships and their crew. And, as he is leaving the encounter, Ishmael begins to have his own thoughts along those suggested by Elijah. Ishmael also suspects foul play.

But the thought is disturbing, particularly since Queequeg had not noticed the men. Ishmael would have to stand on his own and admit to Queequeg, who also thought Elijah a quack, that the sailor might be telling the truth about Ahab. "But I beat the thing down" (chapter 21). He returns to consider a sleeping sailor, whose body is clearly before both himself and Queequeg.

Within all the seeming mystery of Elijah and the misty weather, this fact should be clear to my readers, namely, Ahab is sneaking a private crew aboard the Pequod. *Ahab's action is deceitful.* If Ishmael is responsible for his own leaning toward omens, we will see later, that Ahab produces numerous ones of his own to obscure the evil of his monomania. It was not only the crew of the Pequod who suffered from his fixation but all those who had invested time and money in the voyage—such as "Aunt Charity," who, in chapter 20, "All Astir," helped outfit the boat and who "owned a score or two of well-saved dollars" invested in the success of the voyage.

Indeed, I devote considerable space to the outfitting of the ship by the two principal owners, captains Bildad and Peleg: "As for Bildad, he carried about with him a long list of the articles needed .... Every once in a while Peleg came running out ... roaring at the men...." After all this care and attention, I make Ahab's deceit to have consequences beyond that of deaths of his crew, namely, the loss of the Pequod itself. And yet, until that final destruction and the mysterious escape of Ishmael, my tale is a whaling voyage in search of those magnificent leviathans, sperm whales, traveling and nesting in the oceans of our world. The destruction of the Pequod, *after its lone voyage*, cannot dissolve that tale of whalers and whales, any more than death can dissolve the meaning of life. And then too, my readers must keep recalling, that I was paid to write and had to satisfy both those who paid me and my conscience—*this latter by the twice-told nature of my tale yet to unfold!*

Queequeg

'Cap'ain, you see him small drop tar on water dere?' ... and struck the glistening tar spot out of sight. (chapter 18, "His Mark")

While Ishmael receives only a three-hundredth lay (the percentage from the profit of the voyage), Queequeg, to repeat, is to receive a ninetieth lay. And the partial owner, Peleg remarks, "and that's more than ever was given a harpooner yet out of Nantucket." Queequeg has remarkable control over his body, some of which disturbs Ishmael. In chapter 17, "The Ramadan," his religious fast is so prolonged that Ishmael is moved to lecture him about the dangers of the extremes in any religion. But Queequeg just smiles, "he thought he knew a good deal more about the true religion than I did." Earlier, in chapter 12, "Biographical," Ishmael discovered that, "In Queequeg's ambitious soul, lurked a strong desire to see something more of Christendom than a specimen whaler or two." Queequeg goes to great lengths to discover what civilization has to offer, and we are to suspect that the inconsistencies of killing whales for money are not lost on him.

Queequeg is of royal blood, and he understood the dangers of the hunt. And, although he temporarily lets himself be taken in by Ahab's monomania, the real hunt for him and the other "normal" harpooners, mates, and sailors was pitting one's self against the whale. This "honest" contest takes place several times in my *Moby-Dick*, before the eventful last encounter with the white whale.

On the other hand, despite the actual disaster—as I actually describe, with enough warnings in the text to think twice about what truly happened to the Pequod—there exists in my *Moby-Dick* the recurrent theme that the tragedy might have been averted. The "gams" give us alternative views of the way captains may relate not only to their crew but to the danger of catching whales. Recall that in chapter 100, "The Pequod Meets the Samuel Enderby of London," the captain of the ship, Boomer, had his arm chewed off by Moby Dick. But, not only did he have no thought of revenge on the whale, he joked about it with his surgeon and seemed almost proud of his encounter. When he becomes aware of Ahab's vengeful search, Boomer thinks Ahab is mad, and whispers to Fedallah, who had boarded the Enderby with Ahab, "Is your Captain crazy?"

Indeed, it gradually becomes clear to all on board the Pequod that Ahab has become fixed in his evil, with little hope of changing. Most of the crew, despite having been befuddled by Ahab's theatrics, such as appearing to control the lightening and correcting the ship's compass, now begin to suspect that Ahab's madness may lead them to destruction. But, they never allow this self-knowledge to lead to revolt. True, they are to blame for their weakness; and yet my Ishmael, who himself experienced the terror of Ahab's monomania and did nothing, provides all with a partial excuse for not standing up to Ahab, namely, the power of superstition, itself strengthened with portents and prophecies.

Still, while the killing of sperm whales is a business, it is also, for the men on board, a deed in which they pit their own lives against the great whales. This is not the age of guns that shoot forth harpoons safely from protected large ships—I have seen it once or twice—but the chasing of whales in small boats in which every life is in mortal danger. Thus, after leaving on Christmas day for the great oceans, I give my readers the first eulogies of whaling. "I am all anxiety to convince ye, ye landsmen, of the injustice hereby done to us hunters of whales" (chapter 24, "The Advocate").

American Whaling Men

Chapters 24, "The Advocate," and chapter 25, "Postscript," are another beginning to my eulogy of American whaling men, a eulogy that is to run throughout my *Moby-Dick*. The text is fairly straight-forward. American whaling men, "eighteen thousand men," is larger than the combined whalers of all the other nations of the world. The expense is huge but so is the profit. Moreover, indirectly, American whalers, in their great discoveries and contacts, have preceded all the "Cooks" of the world. They have indirectly paved the way for peaceful relations with natives, who might otherwise have been abused, as was true of many of the Spanish, who sought to possess their lands and labor rather than trade with them.

Nevertheless, an even more general purpose pervades these eulogies, that I give, namely, the praise of America, when compared to the other countries of the word, particularly, England. A cannibal, like Queequeg, if he is a good harpooner, can be the most important and honored member of the crew, second only to the mates and captain. Thus, in chapters 26 and 27, "Knights and Squires," the harpooners are referred to as the "squires" of the usual three mates of a whaling ship. Indeed, the harpooner fulfills many of the most dangerous tasks associated with whaling: "The dangers of a harpooner's life arise not from the abuses of some Captain punishing a man because of his race, but from the tasks inherent in their position. For the most part, they were compensated, both in prestige and in obtaining a greater share of the profits than the common sailors."

The people of Nantucket were broad-minded, but there was also the realization that a savage harpooner was, perhaps, the best source of profit, not by exploitation but because they were less fearful of death. When, after signing Queequeg as harpooner for the Pequod, the pious Bildad tries to convert Queequeg, I have Peleg to shout, "Pious harpooners never make good voyagers—it takes the shark out of 'em" (chapter 19, "His Mark"). Also, one must admit that there seems to be no record of a cannibal owning a home and becoming an accepted member of Nantucket society. More to the point of my tale itself, we have seen that, while the people of Nantucket accepted cannibals, they were taken back by the friendliness between Ishmael and Queequeg (chapter 13, "Wheelbarrow").

Aside from the content of these chapters, their place in my narrative is interesting. You would not be wrong in seeing an almost consistent rhythmic pause between discussing Ahab on the one hand and praising whaling and whales on the other. All this continues until the last three chapters of this 135-chapter book. Indeed, as we approach the "climax" of Ahab's three-day attempt to kill Moby Dick, this chase is preceded by an almost continuous discussion of whales and whaling that extends from chapter 60 to chapter 98—an almost forty-chapter eulogy that effectively overshadows the mad chase of the last three chapters.

I hope that my text itself of *Moby*-Dick will reveal three waves of gathering momentum, each with their appropriate reflections—one, about Ishmael, who is writing this tale years after it occurred, another about whaling and whales, which grows to its greatest height in the thirty-nine chapters, 60–98, and a third about Ahab, not merely as a captain of a whaling ship but as this particular captain hunting Moby Dick. All three currents met and intertwine throughout the narrative; but, the whaling adventure absorbs them all and I hope the reader will grasp this. To review: Ishmael repeatedly reminds us that he is writing this narrative after taking many other voyages. These later "wild" voyages have been about whales. Ahab is, after all, searching to kill a whale, Moby Dick, and while his distinctive presence in that role is only a relatively small part of the long three-year voyage, it is, nevertheless, important. Ishmael, Ahab, and Whales—these three textual waves interrupt each other so that when a reader tends to become overly absorbed in one tale, we are reminded that the other two exist. Yes, all three tales exist; but the tale of the leviathan ultimately absorbs the tales of Ishmael and Ahab—for if Ishmael escapes from the Pequod, so does the giant white sperm whale, Moby Dick, on his own power, without the aid of a life-raft or a rescuing ship.

Introducing Ahab

'*He is Ahab* boy; and Ahab of old, thou knowest, was a crowned king.'"
"'And [replied Ishmael] a very vile one.' (chapter 16, "The Ship")

Ishmael has signed papers and thus committed himself to sail on the Pequod, but he still wishes to see the captain of his ship. Captain Peleg asks to what purpose, and, we may wonder how many ordinary sailors would make such a request. No doubt, it should be clear to my readers that if they were curious about the person who was to be their Captain they would have inquired about him before boarding. But one wonders what purpose such enquiries would have had. When Ishmael responds to Elijah's veiled warnings about Ahab, and says that he was told that Ahab is a good whale-hunter, and "a good captain to his crew," Elijah responds, "That's true, that's true." Indeed, could Ahab have had a worse reputation than the pious Bildad, who, when he was an active captain, drove his men so hard that they had to be carried ashore—weak from overwork and undernourishment. And then, we must recall that Ahab had made many successful voyages; there was no reason to suspect that this would not be another such.

But, again, my Ishmael is writing this tale after he had written about the destruction of the Pequod, and he knows from the beginning of his writing that this voyage will be different. We are back once again to the first chapter, "Loomings," but now, perhaps, with an invitation to reflect upon the biblical names. For Ishmael was the son of Abraham, through his second wife, Hagar—given to him by his wife Sarah, when she had given up hope of having a son. Later, Sarah will surprisingly give birth to Isaac, who will father the "chosen people." My narrator is not named "Isaac," but "Ishmael." Well, "Isaac" would not have worked; it lacks the needed rhythm—and, the needed ambiguity. For our Ishmael is more "metaphysical," than clearly religious. The name "Ahab," however, seems more clearly evil, but Peleg tells Ishmael, "'Twas a foolish, ignorant whim of his crazy, widowed mother" (chapter16, "The Ship"). "Ishmael," "Elijah," "Jonah," "Ahab"—my tale is filled with biblical references. But then my sailors are, for the most part, fighting Quakers, against killing; but somehow whales do not count. "This world pays dividends," Ishmael reflects in chapter 16.

I could, I suppose, expand upon the naming of Ahab, his role in the Bible compared to his role in *Moby-Dick*. But my readers will not all be Quakers, and so perhaps I should at least mention that King Ahab had four encounters with prophets, the first and last with Elijah. But, it should be clear to my readers that, whatever the biblical reference, here, in my *Moby-Dick*, Ahab does not start so fixed in evil that there is no hope for his change. Ahab slowly allows himself to fall deeper into evil, embracing it when he refuses to help find the lost son of the Captain of the Rachel.

And, you may ask, what do I mean by "evil?" I think there is a commonsense meaning of evil that touches all of us, namely, giving such a priority to one's own needs that harm is done needlessly to others. If Ahab wishes to devote his last days to killing Moby Dick, then he should have hired his own boat. But then all would have known him to be crazy, and he would have lost his reputation. Ahab plans to do his evil away from prying eyes, and at the expense of others. The result is that the battle aboard the Pequod between Ahab and his mates becomes as dangerous as that between the whalers and the sperm whales that they chase. The Pequod is lost not so much because of Ahab's insane search after Moby Dick, but because of the failure of his mates, particularly Starbuck, to stand against him. This is the real tragedy within my *Moby-Dick*.

Reference

We frequently hinder and redirect the clarifying light of our own reflections upon our behavior: We choose not to know, and this choice is effective: *We do not know that which we do not wish to know.* I believe that we now have a partial grasp of why I resort to omens and portents, namely, because when we cooperate in evil and refuse to face our own deeds, we begin to see the workings of fate, or of some force beyond that of our own individual powers leading us to evil and preventing us from doing good. Starbuck is inclined to believe in portents, and he is also too weak to oppose Ahab. Yet, my Ishmael explicitly hesitates to blame him. Even Queequeg seems to go along with Ahab's mad desire for revenge. And now Ishmael begins to see that he also should have been more awake to the real danger signs and less prone to give weight to omens.

Are these omens Melville's (through Ishmael) way of allowing us room to excuse the moral weakness of those on board the Pequod to revolt against Ahab? And, if so, have we solved the mystery of the persistent presence of omens within the tale of my *Moby-Dick*? No, but we have made a start

CHAPTER 7

Ahab as Captain and Ahab as Ahab

Abstract The chapter heading clearly imitates a play, and this form is to be repeated several times, more or less clearly, for example, in chapters 36, "The Quarter Deck: Ahab and All," and 121, "Midnight, on the Forecastle." It is as if Melville was informing us that if we want a play such as Shakespeare wrote for his time, he is here giving us—in passing—a play or two just as good for our times.

Keywords Shakespeare • Ahab • Stubb • American Whaling Ship

As the ship sails, we meet most of crew in their order of importance, with the exception of Ahab. In chapter 28, "Ahab," the Captain finally appears on deck. There is no sign of sickness about him; he appears as if cut from stone, face grim with a strange white scar running down one side disappearing into his clothes. So strong and domineering does his broad body appear that it is only as a second glance that one notices his whale-bone leg resting on a small notch made to support it. Thus, I have Ahab introduced to my readers in these fifteen chapters that I am considering with you now as a group. Before I turn to the content of the chapters, I wish to draw

(Chapters 26–40 of my Moby-Dick: "Enter Ahab, to Him, Stubb.")

J. S. Catalano, *In Pursuit of Moby-Dick*,
https://doi.org/10.1007/978-3-031-40357-6_7

attention to the strange way some of the chapters are introduced. They are meant to recall the unique parenthetical remarks that introduce the "Etymology" and the "Extracts," although their function is here different. Let us consider the group. Of these fifteen chapters, 26–40, nine are typical in form, even if some of the vocabulary may seem out of the ordinary. The terms in chapters 26 and 27, "Knights and Squires," are indeed part of our ordinary vocabulary. Chapter 28, "Ahab" as well as chapters 30, "The Pipe," and 31, "Queen Mab," are also typical, although the immediate clarity of the last, "Queen Mab," is not—she is the midwife of dreams. Then too, the forms of chapter 32, "Cetology," and chapter 33, "The Specksynder," are again typical, even if the vocabulary is technical. Finally, chapter 34, "The Cabin-Table," and chapter 35, "The Mast-Head," are clear both in the form of the chapter headings and in their meanings.

I expect that my readers will experience a real jolt, however, in the form of chapter 29, "Enter Ahab; to Him, Stubb," which appears to be a heading taken from a play. This sort of director's comment continues also in the five remaining chapters—36, "The Quarter Deck," 37, "Sunset," 38, "Dusk, "39, "First-Night Watch," and 40, "Midnight, Forecastle." And these forms return in chapters 119–122. Why do I break the traditional narrative form in this way?

Yes, I was re-reading Shakespeare, and yes it seemed like a good notion to emphasize my Ahab as a tragic figure. But, if the truth be known, I do not recall what I had in mind. Who knows? Someone may have knocked on his door, and as I said "Enter," I said to himself, "Why that is a nice way to begin this chapter." Outside knowledge about a text is useful, but there is still the text itself. And, while each reader will come to a text with a different background, a good text speaks to all. My *Moby-Dick* is a good text, and, as such, if one is patient, it will itself explain these atypical narrative forms.

I think that everything becomes clear in chapter 33, "The Specksynder." Here, I explain that the term refers specifically to the chief harpooner, who formally had almost total control of the ship while hunting whales. American whaling ships still give special dignity to the harpooners, who eat apart from the crew, with the captain and his mates, although at different sittings. After making this observation, my Ishmael then comments on the long voyage of an American whaling ship—more than three years—and how, during this time, the captain, *any captain*, is king and lord of all on board. We must recall that Ishmael has had many other voyages; he has seen captains strut the deck as if wearing purple, almost asking the crew, including the mates, to remove their shoes before approaching him.

Thus, the inherent drama, natural to a play, emerges, almost forced out of the text. If the mates are knights with squires, all are subject to the captain, their king. Knights, Squires, Kings; the form of a play announces this royalty, a royalty without robes, but a royalty nonetheless. Ahab may pound the deck at will with his peg leg; but, when Ishmael compares Ahab to other captains, Ahab's distinctiveness radiates from within rather than coming from any kingly posturing. Ishmael concludes chapter 33: "Oh, Ahab! what shall be grand in thee, it must needs be plucked at from the skies, and dived for in the deep, and featured in the unbodied air!"

The dramatic form of play-writing returns in the four chapters, 119–122, "The Candles," "The Deck," "Midnight, on the Forecastle," and "Midnight, Aloft." This later is the shortest chapter in the book, a small paragraph, a simple humorous remark, made by the harpooner, Tashtego:

> Um, um, um. Stop that thunder! Plenty too much thunder up here. What's the use of thunder! Um, um, um. We don't want thunder; give us a glass of rum. Um, um, um!

Simple, but effective in dampening all the preceding drama in which Ahab, seemingly controlling the lighting, stirs up the crew again to chase and kill Moby Dick. Here the drama comes more from the sea and the air, the constant companions of these sailors. In both these groups of chapters, the ones we are considering here and those in 119–122, a storm turns the crew's attention away from Ahab to the safety of the ship–weather, water, and whales, the Pequod's true constant companions.

Starbuck

"The chief mate of the Pequod was Starbuck" (chapter 26, "Knights and Squires"). The shadows of omens chase Starbuck throughout the story, even to its end. In chapter 133, "The Chase—First Day," Starbuck says to Ahab, "tis a solemn sight; an omen." And Ahab replies, "Omen? Omen?—the dictionary! If the gods think to speak outright to man, they will honorably speak outright." Ahab's evil genius—or rather, genius become evil—nevertheless, remains genius; he judges his men rightly. They are too weak and superstitious not to obey him. But then, Starbuck is, at heart, a fine man, honest, and true. He may not have been able to stand up to Ahab but then Ahab was no ordinary challenge. My Ishmael tells us explicitly that he hesitates to blame Starbuck: "But, were the coming narrative to reveal, in any instance, the complete abasement of poor Starbuck's fortitude, scarce might I have the heart to write it" (chapter 26, "Knights and Squires").

And yet, I make it clear that Ahab has his own superstition, his belief in his own harpooner, Fedallah, who has predicted that Ahab would see him dead before he, Ahab, could die. Indeed, Ahab is well aware that he is allowing himself to become befuddled by omens: And, omens follow throughout this tale, for what Starbuck encounters is not merely the danger of whaling, as great as this is; but a clash between good and evil, which itself bring forth the omens, to obscure this very clash. Here in my novel, this combat occurs on two levels at the same time. We have, on the one hand, the daily tasks of the sailors in face of the life and death pursuit after sperm whales. Regardless of Ahab's madness, these tasks can be done with justice and charity or with a selfish spirit that seeps within the souls of all on board. The nine gams—the meeting of two whaling ships—given in the tale reveal that good captains have happy ships, while evil ones have sad ships. There are two extremes given to us: the fourth and the sixth gam—the first, the stupid captain of the Rose-Bud, and the latter, the joyous captain of the Bachelor.

But, in a sense, the Pequod stands midway between these two extremes. For, once Ahab has discovered that the captains have not sighted Moby Dick, or have sighted the whale some time ago, he then leaves the running of the ship to his mates. On this level, the tension between Ahab and the crew is in the background, the dangers of whaling itself take over the tale, and here the mates are competent and good. There are disasters, but not arising from carelessness or selfishness.

On the other hand, the broader conflict does exist, namely, the struggle to keep their minds on their daily tasks while they know that their captain and god has vowed to take vengeance on the giant white sperm whale, Moby Dick, regardless of the consequences. Again, each of the gams reminds the sailors of Ahab's monomania; but, it is not until the gam with the Rachel that Ahab puts the chase after Moby Dick before all else, even the possibility of saving a life. To repeat my earlier observation, it is not until chapter 128, "The Pequod Meets the Rachel," custom and moral right would have demanded that Ahab join its captain, Gardner, in his search for his son, now lost at sea. It is at this time that Ahab's maniacal search for the one whale, Moby Dick, leads to a clear act of injustice as distinct from theatrics. Gardner has also sighted Moby Dick, and Ahab stresses his rejection "in a voice that prolongingly molded every word—'Captain Gardner, I will not do it.'"

The complexity of this fight between good and evil is real, involving as it does human freedom. We are made aware that had Starbuck stood his ground—when, for example, he replied to Ahab that they did not go on this journey to seek his vengeance on a blind brute but to serve the owners in catching as many sperm whales as possible—the entire tale would have taken a different turn. Nevertheless, this possibility, namely, that all might be different than it was, is itself part of the tale—it haunts the tale.

My Ishmael, however, knows that the mates will not seize control of the ship. Thus, in describing the mates in these early chapters, Ishmael is retrospectively searching for reasons that might justify their weakness. Starbuck is a truly fine man, a moral person who would not be afraid to risk all for justice; but he is superstitious. And, Ishmael reflects, that this superstition seems to arise more from intelligence than ignorance. There are two kinds of bravery, one that confronts the endangering forces of nature, the other that faces the evil in others. Starbuck fears Ahab more than Moby Dick.

Stubb

"Stubb was the second mate … Good-humored" (chapter 27, "Knights and Squires"). In chapter 29, "Enter Ahab; to Him, Stubb"—again, I repeat what I mentioned earlier, for the context is different—the second mate suggests to Ahab that he might wish to muffle the sound of his whale-bone leg while passing the deck at night, so that the men might be able to sleep. Ahab rebuffs him, calling him a "dog" and then "ten times a donkey, and a mule, and an ass." Stubb replies that he is not used to being spoken to in that way; but then Ahab advanced "upon him with such overbearing terrors in his aspect, that Stubb involuntarily retreated."

A captain has the right to give normal orders to his mates, but not to abuse them in this verbal way. Stubb was an easy-going second mate, and he took the abuse; he should not have done so. Was Ahab going to strike him? That direct confrontation was not in Ahab's character, and, besides, it would have been illegal. Stubb should have insisted upon an apology; instead he backed down.

Stubb backs down from Ahab's gaze, and he then dreams. In chapter 31, "Queen Mab," Stubb dreams that Ahab had kicked him with his ivory leg; but he thinks to himself that, after all, it was only an inanimate object that kicked him. A slap in the face by real hands would have been worse. Stubb clearly cannot face the shame of the insults he received from Ahab, and he attempts to hide them in a dream, a dream that must be obvious even to him.

I have this strange chapter to end, leaving the dream world, and with Ahab shouting to his men to keep an eye out for a white whale—"split your lungs for him!" And Stubb wisely wonders why all the fuss about a white whale, when their business is catching as many whales as possible. And, as if interpreting our wonder and leading it away from Ahab's search for Moby Dick, Ishmael follows this chapter with a discussion of whales in general, chapter 32, "Cetology," which Webster simply defines as a branch of zoology dealing with whales. I have already discussed above that I consider the comparison of whales with books to be fairly accurate. Then too, it is very much in line with the constant reminder of the value of books and the mystery of writing that begins with the preliminary sections, "Etymology" and "Extracts," and continues throughout the narrative: The written word "whale" is as much a wonder as a real whale.

Flask

The third mate is Flask. He is described as one who chases after whales as if they were a species of overgrown mice. If Starbuck was serious and Stubb easy-going, Flask was short, stout, and solid, like a foundation that could withstand seas and whales. "Now these three mates—Starbuck, Stubb, and Flask, were momentous men" (chapter 27, "Knights and Squires"). And, like the knights of old, they each had their squires, namely, their chosen harpooners. Queequeg was the squire of Starbuck, Tashtego, a native American Indian, was the harpooner for Stubb, and, the short Flask had the tall, imperial negro from Africa, Daggoo. Ishmael tells us that whenever the sight of gloomy Ahab would shake him, he would tell himself that nothing could go wrong with these three mates.

And nothing should have gone wrong, except, of course, the natural mortal dangers of the quest itself. For, we must keep in mind that the Pequod might have been destroyed by a sperm whale *in the normal course of the adventure itself.* It is not the destruction of the Pequod by the sperm whale, Moby Dick, that is extraordinary; but, rather, that the chase after him was done for vengeance, and long after it should have been given up as a hopeless encounter.

On the other hand, before the chase of the last three days, Ahab had abided his time. He did not refuse the search for other whales, and as much as possible, he kept the Pequod running on its normal course, even with his theatrics: "For be a man's intellectual superiority what it will, it can never assume the practical, available supremacy over other men, without the aid of some sort of external arts and entrenchments, always, in themselves, more or less paltry and base" (chapter 33, "The Specksynder").

Thus, in chapter 36, "The Quarter Deck" I have Ahab rouse the men with the adventure of seeking Moby Dick, nailing a golden doubloon dramatically to the main mast to be given to the first to sight the whale; he then passes drink to all as he has them swear death to Moby Dick. And, yet, I wish my readers to see these as tricks used by anyone attempting to use others. To understand my Ahab, I would have my Ishmael go deeper into Ahab's soul. Why is he here and not retired to be an owner as Peleg and Bildad did? Is he afraid of becoming old? And he has a young wife and child; or is that part of what keeps him at sea?

Ahab as Ahab and Ahab as Captain

"I came here to hunt whale, not my commander's vengeance" (chapter 36, "Quarter-Deck"). Starbuck rightly begins to stand up against Ahab, who then attempts to make it seem that the search for profit is ignoble. But business is neither good nor bad; it is simply part of the machinery of life, and Ahab is well aware of that simple truth. I would have my readers accept the Pequod's whaling voyage for what it was supposed to be, namely, a good-faith enterprise, sponsored by those who were willing to gamble their money to procure goods needed at home. That perfect justice may not have been the rule in paying wages can be granted, without condemning the enterprise itself. Besides, Ahab exaggerates his wound from Moby Dick: "I'd strike the sun if it insulted me." But, is sunburn an insult given to man by the sun, and how will one strike back?

And, yet, my Ishmael suspects that Ahab's monomania has deeper roots than simple revenge on Moby Dick for taking off his leg. He has said, to repeat, "Oh, Ahab! What shall be grand in thee, it must needs be plucked at from the skies." And then in the next chapter, Ishmael remarks about Ahab that he almost wants someone to stand up against him, "I thought to find one stubborn" (chapter 37, "Sunset").

Have I provided enough texts to go deeper into my Ahab? Let us begin with the obvious. He is a captain of a whaling ship, and every captain is a king of sorts. He has his "humanities," as Captain Peleg remarked to Ishmael in chapter 16, "The Ship"; for he has a young resigned wife, not "three voyages wedded," and a child. One would think that this marriage would give Ahab a new spark on life. Of course these voyages are three years long, and he was thus married at least nine or ten years ago. That is, Ahab was wedded two voyages before he had his leg torn from him by Moby Dick—this is his first voyage since that incident. What must his life with his young wife have been like, with his whale-leg? She is described as resigned, and accidents had to be expected by the wives of whalers. But was Ahab resigned to live with a resigned wife?

To continue with what is obvious, Ahab sees himself fading; he is getting old: "Oh! time was, when as the sunrise nobly spurred me, so the sunset soothed. No more" (chapter 37, "Sunset"). On the other hand, we are probably here concerned with more than mere age; but, rather, an inability to enjoy the simple pleasures of life. In chapter 30, "The Pipe," Ahab thinks to himself, "Oh, my pipe! Hard must it go with me if thy charm be gone!" And yet, it is clear to Ahab that the pleasure of smoking has vanished not because of age or illness, but because he sees his purpose as too weighty to allow him such simple pleasures: "What business have I with this pipe?" If the dawning of old age is not all that weighs on Ahab, it is hard to picture a young Ahab willing to risk his life in a mad pursuit after a whale that almost killed him.

Also, Ishmael makes it clear that we are not to give too much weight to the mere fact that Ahab lost his leg through his encounter with the sperm whale, Moby Dick. Again, I provide a gam to highlight the distinctive character of Ahab, and, although already mentioned, it deserves repeating. In chapter 100, "The Pequod Meets the Samuel Enderby of London," the captain of the ship, Boomer, has lost an arm through his encounter with Moby Dick, and Ahab goes aboard to listen to Captain Boomer recount the incident.

Never did two men react so differently to similar events. For Boomer, it was an exciting incident, not only for himself but for his surgeon, Dr. Bunger. Ishmael seems to enjoy detailing the conversations between these two, as Ahab becomes almost mad with impatience to leave once he discovers that Boomer did not kill Moby Dick. It is clear that Ahab does not wish the death of Moby Dick, except by anyone but himself. It is even more evident that whatever harm Moby Dick did to Ahab, another captain would have been glad to have escaped alive and would never dream of seeking to kill him for revenge.

All of the above are merely meant as suggested reasons, but not causes of Ahab's maniacal pursuit after Moby Dick, and the possible deeper roots of his evil. In the final analysis, Ahab's evil is an act of freedom, whose depths will become, if not clearer to us as we proceed, then, perhaps, more understandable. Here, let me simply suggest, that, perhaps, the true root of Ahab's evil is the fear of ending his life, "all too human."

Introducing Pip

AZORES SAILOR, Dancing:

> Go it, Pip! Bang it, bell-boy! Ring it, dig it, sting it, quig it, bell-boy! Make fire-flies; break the jinglers!. (chapter 40, "Midnight, Forecastle")

We have met the mates—Starbuck, Stubb, and Flask (the "knights"), and we have met the harpooners—Queequeg, Tashtego, and Daggoo (their "squires"). We have met their Captain, Ahab, and his sneaked-aboard harpooner, Fedallah. But we have not been formally introduced to the rest of the crew, and this is done in chapter 40, "Midnight, Forecastle," which continues the dramatic play-form introduced in chapter 29, "Enter Ahab; to Him, Stubb."

Most of the sailors are happy with the drink dramatically given to them by Ahab, when he managed to get them riled up for his private chase after Moby Dick. Starbuck's objections to the craziness of this private vengeance almost broke through Ahab's semi-hypnotic affect on his crew; but he too backs down. He would have had to stand up against the harpooners, including Queequeg, who seems game for the fight, and to the crew who wished for the gold doubloon promised to the first man to spot Moby Dick. If Starbuck had only repeated his just observation that they did not sail to help him with his private vengeance, the disaster of the Pequod—as here recorded by my Ishmael, with my reservations to come—might have been adverted, at least as caused by Moby Dick.

Still, it should be interesting that Ahab feels it necessary to pass drinks all around, "Drink and pass!" (chapter 36, "Quarter Deck"). Is he afraid that the mood he has induced in the crew will pass too quickly? In any case, the crew has drunk deeply, and in chapter 40, dear readers, we are introduced to their varied background through episodic play-like interactions. Sailors sing and dance. They are from the States—Nantucket and Long Island—but also from all over the world, France, Spain, Malta, Portugal, and so forth. If there is a center to the merrymaking, it is the cabin-boy, Pip.

Pip, the cabin-boy, becomes a central figure as my narrative continues. He plays a tambourine, comments upon events, and he has the run of the ship. Later, in chapter 93, "The Castaway," he falls from a boat and is almost drowned, when he has to replace one of the members of Stubb's crew as they are chasing a whale. Stubb thinks that the boat behind him will pick Pip up (this usually happens), but the other boats have also sighted whales. Ishmael comments in the same chapter, "Man is a money-making animal, whose propensity too often interferes with his benevolence." Pip is left to his own resources, he almost drowns, but he is picked up by the Pequod itself.

From the time of his narrow escape from deep waters, Pip is no longer joyous and lucid; he seems to have lost his reason. Yet, in a strange way, Pip becomes close to Ahab, and Ahab to Pip. In chapter, 129, "The Cabin: Ahab and Pip," Pip says to Ahab, "Ye have not a whole body, sir; do ye but use poor me for your one lost leg; only tread upon me, sir; I ask no more, so I remain part of ye." Ahab is moved, almost seeing his own madness reflected in Pip; but he sticks to his purpose of chasing Moby Dick. He leaves his cabin, but requires Pip to stay, "God for ever save thee, let what will befall." But, of course, in the ordinary understanding of the narrative, Pip dies with all the crew—except Ishmael.

Within all this drama, I give a whole chapter, of decent length, on the mast head of an American whaling ship, chapter 35, "The Mast-Head." There are no comfortable "crow's-nests" to rest in; but a sailor can put his legs about the riggings of a mast and rest there comfortably, particularly in the calm and warm waters in which they frequently sail. During a three-year voyage, a sailor may spend months there, and if this sailor is contemplative, as Ishmael admits he was, he will be thinking more about the meaning of life, than looking for whales. Looking out over the endless ocean, one may seem to be dissolved into it, mindless of whales and men. If one is not careful, a slip may dissolve you in that sea. Ishmael, who no doubt came very close to falling, warns, "Heed it well, ye Pantheists!"

CHAPTER 8

Ahab and Moby Dick

Abstract What is important here is the distinction between Ahab as captain of a whaling ship and Ahab in his mad pursuit after the whale that took off half his leg and which he names Moby Dick. Given this distinction, we must keep in mind that every captain has distinct personal characteristics. We must be careful to separate these from the monomania which can be said to be fairly unique to him. Indeed, the chapters devoted to Ahab as he is a typical captain of an American whaling ship far outnumber those detailing his mad pursuit after Moby Dick. As a typical captain, even with personal characteristics, he grants the general running of the ship to his mates and harpooners.

Keywords Ahab • Moby Dick • Pequod • Mates • Harpooners

> The White Whale swam before him as the monomaniac incarnation of all those malicious agencies which some deep men feel eating in them, till they are left living on with half a heart and half a lung. (chapter 41, "Moby Dick")

I have written a great deal, hoping that the text of this book will be so clear that one blinks; and in the time of those blinks, one suspects the

(Chapters 41–52 of my Moby-Dick)

J. S. Catalano, *In Pursuit of Moby-Dick*,
https://doi.org/10.1007/978-3-031-40357-6_8

lurking presence of one's own deep and hidden notions. Thus, although we might rightly refer to my book as "multileveled," we would also have to acknowledge that, most of what is ordinarily merely suggested in a text, is here given to us clearly, as part of my Ishmael's own reflections. Let us, then, reconsider, with Ishmael as our guide, the possible beginnings of Ahab's monomania—first, however, we must again carefully place it within the more ordinary dangers of whaling.

Ahab had gone in pursuit of Moby Dick; but so, indeed, had a few other captains of whaling ships. None had been able to kill the giant sperm whale; rather, they left the encounter with great loss to themselves and their ships. Nevertheless, after the chase, they let the matter rest there. True, as captains of American whaling ships, it was their job to hunt sperm whales, and, some have chased exceptionally large sperm whales, each of which had proper names. There was "Timor Jack," "New Zealand Tom," "Morquan," and "Don Miguel." In regard to the other incidents of chasing giant sperm whales, for example, Timor Jack, I hope that you can surmise from the tone of Ishmael discussion, in chapter 45, "The Affidavit," that the owners and the crew were aware of the nature of the chase from the beginning. There was the fun of the chase—as dangerous as that might be—a bit of glory, a tale to be told to other sailors; and, for the owners and all alike, oil and blubber.

Further, strange as it might seem, even after affecting an escape with a harpoon deep inside it, a whale has been known to swim away and to have been later killed by the same hand that inflicted the first wound. In short, the remarkableness of Ahab's chase after Moby Dick does not consist in the brute facts of the chase as such. Whaling ships, particularly American whaling ships that almost alone hunt the giant sperm whale, are involved in an exciting and dangerous adventure. Sailors perish in pursuit of blubber and oil, and few at home care or are even notified of their deaths.

Nevertheless, granting all the other chases after giant sperm whales, it would be considered lunacy to endanger unnecessarily both themselves and their crew in a mad personal pursuit after a particular whale for no other reason than stupid vengeance. And, if one adds to this, a question of sneaking a special crew on board and of getting everyone riled with dramatics and drink, the Pequod's chase after Moby Dick is insane, and it would be considered such by any whaling standard. But we are still only approaching the root of the insanity that directed this mad chase.

Every chase after a sperm whale has crazy moments. I see to it in my writing that my Ishmael continually reminds us that whaling is a dangerous profession in which life and limb are usually at risk. In my chapter 48, "The First Lowering," and in the following chapter, "The Hyena"—I will return below to consider the title of the chapter—we are given a description of how even the prudent Starbuck looses his boat in pursuit of a whale, manages to get back on his boat, only to be almost cut in half by the Pequod, which did not see his boat in the storm. In this incident, Ishmael and Queequeg were part of Starbuck's crew; Ishmael was rowing and Queequeg was Starbuck's harpooner. When they are finally rescued, Ishmael asks, "Queequeg, my fine friend, does this sort of thing often happen?" (chapter 48, "The Hyena"). When Ishmael is informed that it is not uncommon, he then goes down to his bed to make up his will, which, we are told is a common sport among sailors.

Chasing great sperm whales, danger, the possibility of losing one's life, all this is an "ordinary" part of the American whaling adventure of the early to mid-nineteenth century. Finally, before we attempt to properly place the distinctiveness of Ahab's own evil, we must give some thought to the evil of killing whales. Is the killing of whales for profit sane? Is a war blessed by a pope or king sane? To repeat an earlier observation, these killings of sperm whales could be more accurately compared to medieval knights jousting than to war, or to the American Indians, who killed a buffalo because every part of the animal was needed for the necessities of living. Indeed, I give this very comparison in chapter 105, "Does the Whale's Magnitude Diminish?—Will He Perish?" There is this important difference between whales and buffalos, namely, whales cannot be so easily cornered for a mass killing as buffalos, and, further, the task, for the most part, is much greater: "Forty men in one ship hunting the Sperm Whale for forth-eight months think they have done extremely will and thank God, if at last thy carry home the oil of forty fish."

Thus, although Ahab's pursuit after Moby Dick was insane and evil, it was not so crazy that it could not be somewhat placed as part of the natural whaling experience. The harpooners, Queequeg, Tashtego, and Daggoo, were all temporarily taken in by Ahab's monomania, for they saw in it the glory of the chase and not the evil in Ahab's intention. But, now, with all these caveats, we can begin to approach the sources of Ahab's evil.

Ahab's Pain

"It is not probable that this monomania in him took its instant rise at the precise time of his bodily dismemberment" (chapter 41, "Moby Dick"). I believe that our general experience confirms the truth of my above claim in my book. Whatever the causes of an injury, when it occurs, we are usually aware at first only of the injury itself. Thus, I have my Ishmael to come to the conclusion that Ahab's insanity began to take hold of him later, during the voyage home, as he lay in agony. At times, Ahab had to be strapped down, so intense was the pain and so near insanity was he from the pain. Gradually, however, as the ship entered warmer waters, he seemed to gather strength and be more himself. The crew expected that he would show some deep affect from the wound; but his manner seemed overly calm.

Aside from the pain, my Ishmael makes a point of detailing the almost crazy circumstances during which it was inflicted. Ahab had been in charge of his three boats, each was attempting to get the great white whale now first sighted. During the fight, Moby Dick had created such an eddy that all three boats were whirling about, harpoons and oars lost. Ahab then seized a short six-inch blade with which he attempted to kill the whale, a hopeless gesture. Any sailor would have known that one could not reach through the whales blubber with such a short thrust. It was then, during this useless attempt to kill Moby Dick that the white whale turned and tore off Ahab's leg.

If any other sailor had so acted, he would have been reprimanded for his foolishness. No doubt, Ahab was a skilled and dedicated hunter of whales, and his mates probably thought the incident reflected his simple anger that the whale was escaping from them. Still, if Ahab was foolish and crazy in this thrust, it was not the character of his later madness, during which he was, for the most part, in control of himself.

Still, you will note, my Ishmael surmises that the incident is important, for the paragraph describing it in chapter 41 is given to us in the manner of an indirect discourse. Textually, it stands out from the rest of my text in the chapter. That is to say, the attack of Moby Dick on the three boats chasing him and the useless attempt of the captain to kill the whale is given as if it had happened to some other captain, a foolish one. But then we are informed: "That captain was Ahab." If it is not clear exactly what is being suggested here, I wish some mystery to be present, for the reader to consider. Is Ahab secretly angry at himself for his own foolishness? If he had been wounded by Moby Dick during a more proper attempt to kill the whale, he might have felt better about his wound and the consequent pain.

Nevertheless, it is also probably true that his later madness aboard the Pequod after his first wound had other sources; and one of these was no doubt the very passivity of the crew, which "seemed especially picked and packed by some infernal fatality to help him to his monomaniac revenge" (chapter 41, "Moby Dick").

My Ishmael Is Also Taken In

> I, Ishmael, was one of that crew; my shouts had gone up with the rest; my oath had been welded with theirs; and stronger I shouted, and more did I hammer and clinch my oath, because of the dread in my soul. (chapter 41, "Moby Dick")

Every ship is an oasis in the waters of the world, or better an island within which good and evil rebound with each person. With another crew, Ahab's madness might have been checked. But, then, Ishmael was part of that crew, and he admits that he let Ahab's desire to kill Moby Dick for pure revenge seep into his own soul. It is his own weakness that he now wishes to understand—but this is a particularly difficult undertaking for any person. Indeed, after reviewing the possible reasons for Ahab's madness and after making a final effort to understand Ahab's power over himself, Ishmael backs away from the effort to understand Ahab's magnetism—"all this to explain, would be deeper than Ishmael can go."

My Ishmael is very adept in reflections, and a reader would be correct in wondering whether he stops here because the thought is deeper than he can go, or, rather, deeper than he wishes to go. How does one stop reflections, particularly those that might be disturbing to one? Let me explain. Consider with me my chapter 42, "The Whiteness of the Whale." Ishmael writes, "What the white whale was to Ahab, has been hinted; what, at times, he was to me, as yet remains unsaid." The key to the chapter seems clear—Ishmael is trying to explain to us the terror that white sperm whale instilled in *him*. But does this key open the door to the chapter's meaning? For what follows is not a personal reflection on whiteness; but, rather, an attempt to show that whiteness is strange to everyone.

Indeed, dear reader, you would be correct in judging chapter 42, "The Whiteness of the Whale," as strange a chapter as the whiteness it attempts to explain. Ishmael produces an inordinate number of long footnotes, none of them convincing. Then, toward the end of his discussion, he seemingly awakens to what he is doing, and tells us how foolish we must think his attempt: "But thou sayest, methinks this white-lead chapter is but a white flag hung out from a craven soul; thou surrenderest to a hypo, Ishmael." The phrase, "white lead" refers here, perhaps, to the lead-type used in printing and not to the color of the printed word itself. Moreover, the chapter is falsely presented as absolutely necessary to understand the rest of what follows. Ishmael says that he must try to explain why the whiteness of the whale "appalled" him, otherwise "all these chapters might be naught." But this is clearly an exaggeration of a type repeated by Ishmael throughout my narrative, bringing to attention the twice-told nature of my narrative. That is, Ishmael repeatedly states that unless you understand this or that, you will not understand what follows. At times, the claim seems true but at other times it seems an exaggeration—that is to say, my reader has to keep awake. For example, chapter 24, "The Advocate," and chapter 25, "Postscript," have a similar character as chapter 42; namely, both chapters advance arguments which apparently have to be understood or all that follows will be lost on the reader. In chapter 45, "The Affidavit," Ishmael writes, "So far as what there may be of a narrative in this book…"; and what follows in my narrative is a long series of chapters that deal with whales in general—chapters that have little relation to Ahab or the whiteness of Moby Dick—with the words, "It is time to set the world right in this matter." And, in chapter 60, "The Line," "With reference to the whaling scene shortly to be described, as well as for a better understanding of all similar scenes."

That is to say, frequently throughout the narrative, a particular chapter reminds us (and Ishmael) both of the twice-told character of this narrative, and of his own attempt to understand his experience aboard that first remarkable voyage of the Pequod. Yes, something very strange happened on that first voyage years ago, and is it possible now to better understand it?

Most of the other reflections that Ishmael gives us in this narrative are astute, and they reveal a seasoned intellect, nourished with years of reading and sharpened with many other whaling voyages taken after this first voyage on the Pequod. Yet, at other times, such as here, he attempts to present himself, not merely as he was in his first voyage, with all the difficulty inherent in that reflection, face up, before us; but as someone consciously attempting to alter the very narrative he is relating. We are back to the unique twice-told character of this adventure, which began in the first chapter, "Loomings." And we may wonder about this second narrator, "*Who is this Ishmael, and how is he related to Ahab's madness?*"

Trying to Understand Ahab

"This is much; yet Ahab's larger, darker, deeper part remains unhinted" (chapter 41, "Moby Dick"). I think that the above is a conscious exaggeration by Ishmael; for, indeed, he has fairly well described for us Ahab's evil. The purpose of the exaggeration is, again, to give us excuses for the crew's passivity in the face of Ahab's evil. And, yet, Ishmael has admitted to us that there was no excuse. What is sought here by Ishmael is an understanding of weakness in the face of evil, indeed, of the cunning of evil, which Ishmael describes accurately for us. "Human madness is oftentimes a cunning and most feline thing" (chapter 41, "Moby Dick").

Ahab did not lose any of his intelligence in his madness; rather, both became more focused and more hidden; or, in Ishmael's words, no longer an agent but now only an instrument:

> But, as in his narrow-flowing monomania, not one jot of Ahab's broad madness had been left behind; so in that broad madness, not one jot of his great natural intellect had perished. That before living agent, now became the living instrument. (Chapter 41, "Moby Dick")

What precisely is the meaning of this distinction between the intellect as living agent and as a living instrument? You would be right dear reader to see it as the difference between good and bad faith. In what one might term "good faith," we are each joined to our intellect as to our body—the whole (body and intellect)—confronts the world as one principle of action. In our moral dealing with people, we do not pretend that our body or our intellect acts independently from us and that our evil actions are done not by us but by these strange instruments that have taken over our life. We pretend that we do not really understand what this intellect of ours is doing, but we decide to let it have its way. Thus, if self-deception is going on within us, we allow ourselves to believe that it does not originate from us, but from this strange thing within us, our "intellect." And, strangely, we can become temporarily lucid to what we are doing, and yet allow it to continue. If you were thus to elaborate on my text, you would not stray from the text, as indeed should be clear from chapter 44, "The Chart":

> But as the mind does not exist unless leagued with the soul, therefore it must have been that, in Ahab's case, yielding up all his thoughts and fancies to his one supreme purpose; that purpose, by its own sheer inveteracy of will, forced itself against gods and devils into a kind of self-assumed, independent being of its own.

Ahab wills to be controlled by his will and his intellect. He knows that he is surrendering to his monomania, and he wills to remain true to that vow. From the beginning to the end of my narrative, incidents constantly arise in which Ahab is on the verge of letting go of his hatred; but, then, he holds true to it. I have already mentioned that in chapter 132, "The Symphony," Ahab almost yields to the calmness of the seas and to his reflection on his life to alter his course: "Oh, Starbuck! it is a mild, mild wind, and a mild looking sky. On such a day—very much such a sweetness as this—I struck my first whale—a boy harpooner of eighteen!" And, after Ahab's recollections, Starbuck tries to get Ahab to give up his chase after Moby Dick. But, then, Ahab becomes distracted, and Starbuck realizes that he is no longer listening to him, and moves away.

We are thus back to the existence of portents and omens, both as ridiculed by Ahab and yet as he is taken in by them. For as we pretend that our intellect is not under our control—that we, obey *it*—shadowy forces begin to emerge, which seem to have final control over our lives. Indeed, throughout this tale, Ahab's monomania is revealed as the result of free choice and a destiny itself created and sustained by that same free choice. I suggested before that Ahab's real sin is the fear of ending his life in an all-too-human way. It is one thing to wish to live a great life, another to want to redeem a life by seemingly taking the world on one's shoulder. Ishmael perceptibly writes about Ahab:

> He piled upon the whale's white hump the sum of all the general rage and hate felt by his whole race from Adam down; and then, as if his chest had been a mortar, he burst his hot heart's shell upon it. (Chapter 41, "Moby Dick")

That is to say, Ahab would have us believe this of him. But, he was only an old man, whose pride was hurt first by his own foolishness in getting his leg cut off in a senseless attempt to kill Moby Dick and then in his unwillingness to see himself ultimately retired as captains Peleg and Bildad. Or, at least, those human failings must be considered, and they are no small thing.

Free Will and Fate

> Aye, chance, free will, and necessity—no wise incompatible—The straight warp of necessity, not to be swerved from it ultimate course ... free will still free to ply her shuttle. (chapter, 47, "The Mat-Maker")

Earlier in this section, I said I would consider the title of chapter 49, "The Hyena." It represents the deadly laugh of the fates at the power of circumstance and the weakness of men. It would be an error to give this chapter more attention than it deserves, as if the entire narrative calls attention to man's weakness to fight evil. Its limited scope is clearly stated in the opening sentence: "There are certain queer times and occasions ... when a man takes this whole universe for a practical joke...." Ishmael goes on to reflect that as strange as Ahab and Moby Dick seemed, that having almost just lost his life in an ordinary whaling incident, he came to the conclusion that the only way to survive was to have a "free and easy sort of desperado philosophy." Granted, and apparently, our Ishmael took this lesson to heart—recall his "wild adventures" after he survived the Pequod.

Still, this drama of free will and fate is interwoven with the truth of whales and whaling. There exists in my *Moby-Dick; or, The Whale*, a conscious and structured rhythm, which, like the giant waves of the oceans, rises to metaphysical heights only to then return to more natural ordinary events. And, just as the crests and troughs of waves differ according to the violence and calm of the weather or the particular oceans in which they occur, so too the rhythms of metaphysical and poetical reflections seem frequently to vary in *Moby-Dick*. The everyday rhythms in *Moby*-Dick concern those leviathans of the oceans, sperm whales, and also the knights, squires, and kings who pursue them. The presence of Ahab brings these reflections to new heights; but then so does the actual physical presence of the sperm whale itself, both as it is being chased and being cut into. On the other hand, we each in our lives experience similar crests and troughs in our daily pursuits. At times, we awaken to the wonder of everyday events, an unexpected smile, an artifact that works well, but then the business of life tends to dull their splendor. If everything seemed extraordinary and wild, one might read my text as a fantasy rather than as a true reflection of whales and whalers. Thus, in chapter 45, "The Affidavit," Ishmael informs us not to so overrate the extraordinary size of Moby Dick that the encounter with him seems "a monstrous fable, or still worse and more detestable, a hideous and intolerable allegory." For, indeed, *for most of the voyage*, Ahab's monomania is in the background, and what I have written is an adventure tale about whales and the men who hunt them. And yet, there is the two-fold adventure wedded to the one yet to be fully considered.

CHAPTER 9

The Town-Ho's Story and Other Gams

Abstract The notion is explicitly described as a unique meeting of whaling ships, whenever they sight each other. Several gams are described in the narrative, and in all of them, Ahab inquires about Moby Dick, and, when as usually happens, he is given no information, he then retires and lets the meeting continue among his mates. If Ahab in his mad pursuit after Moby Dick had the tight control over his ship that many critics assert, he would have prevented these as wastes of time. But, he does that only in the last two gams, described toward the end of the narrative in chapters 128 and 131, when the other captains have sighted Moby Dick. The Town-Ho's story is a very long description, and it was the only part of the book *Moby-Dick* that was published separately.

Keywords Town-Ho • Gam • The Crotch • Captain Mayhew • Gabriel • Radney • Steelkilt

"Out of the trunk, the branches grow; out of them, the twigs. So, in productive subjects, grow the chapters" (chapter 63, "The Crotch"). I have

(Chapters 53–71 of my Moby-Dick)

J. S. Catalano, *In Pursuit of Moby-Dick*,
https://doi.org/10.1007/978-3-031-40357-6_9

written a very big book and I assume that you do not have a copy, and so let us take another look at all these pages.

1. From the "Etymology" to the "Extracts" to chapter 15, there is no mention of Captain Ahab at all. We thus have 25 pages of preliminary material and 75 pages that comprise the first 15 chapters—100 pages of text with no mention of Ahab. Of course, it is a big book and we have 525 pages or so still to go. With chapter 16, "The Ship," we have the first mention of Pequod and its captain, Ahab, and by chapter 22, "Merry Christmas," the ship is sailing on its three-year whaling voyage. On any such voyage, every ship has mates and harpooners, who are knights and squires in relation to their captain who is their king. Further, the text provides us with brief descriptions of several captains of whaling ships, beginning with the main owners of the Pequod, Peleg, and Bildad, in chapter 16, "The Ship," and continuing throughout the narrative in the nine gams that I will list again below. Again, it is clear that every captain has personal characteristics even as he is in command of his ship. Nevertheless, Ishmael clearly informs us that, a captain's first duty to the owners is to capture as many whales as possible, since the main purpose of the voyage is profit. Thus, referring to Ahab, Ishmael remarks in chapter 41, "Moby Dick":

> Had any of his old acquaintances on shore but half dreamed of what was lurking in him them, how soon would their aghast and righteous souls have wrenched the ship from such a fiendish man!

You would thus be correct to distinguish in my tale between Ahab as he is acting much like any captain of a whaling vessel (Ahab as Captain), and Ahab in his monomaniacal pursuit after the one whale, Moby Dick (Ahab as Ahab).

2. Granting this distinction and being generous in favor of mad Ahab were there is doubt—for example, the long chapter, 54, "The Town-Ho's Story"—we have, *at most*, 50 chapters of my 135-chapter book, which concern Ahab in his mad pursuit after Moby Dick. I will soon give the numbers of these chapters. But, it is important to mention that, if the majority of these fifty chapters were grouped continuously at the end of Melville's tale, then one might make a case that all the preliminary material was leading up to them. The simple truth, however, is that, as we approach the end of my narrative, we have a massive interruption—from chapters 72 to 98, which are an almost continuous description of whales and whaling, with very little to do with Ahab's monomania.

3. Let us begin to be specific. Ahab is formally introduced to us in chapter 28, "Ahab," and in chapter 99, "The Doubloon," he reappears attempting to mesmerize the crew to follow him in his crazy pursuit after Moby Dick. Within those seventy-two intervening chapters, his distinctive presence is felt in twenty-four or maybe twenty-five chapters, but absent from the remaining forty-seven chapters. Furthermore, even after chapter 99, by which we are about three-fourths into the tale, there are still several chapters that are not related to mad Ahab.

4. To be precise, in this 135-chapter book, *not fifty chapters, but, indeed, only forty-seven chapters pertain to Ahab as Ahab*; and this allotment is one that is very generous, namely, chapters 19, 21, 28–31, 36–39, 41, 43–45, 47–48, 50–52, 54, 59, 70–71, 73, 99–100, 106, 108–09, 111, 113, 115, 117–121, 123–26, 129–135. These chapter numbers make it clear that from chapters 55 to 98, only four of these forty-three chapters pertain to Ahab as Ahab.

And, suppose I wished to be even more generous to those who might misunderstand my great book—excessively so, I think—I might then grant all the pages from chapter 99 until the end of the tale to Ahab as Ahab, that is, from pages 470 to 625, to wit, 155 pages. If we then count the pages of the chapters prior to 99 that pertain to Ahab as Ahab we have about 110 additional pages. Nevertheless, we then have only 265 pages of a 650-page narrative that pertain to mad Ahab. And, within these 265, or let us say, 275 pages, we have ample text space for reflections, whether by Ahab, Ishmael, or the mates of the Pequod.

In conclusion, if I wished to write a tale mainly about Ahab's mad pursuit after the white sperm whale, Moby Dick, I could have saved myself at least 350 pages of text, which is to say, more than half my book!

But then, I do not want to go to the other extreme in my rejection of Ahab's monomania as the central moving theme of this great book. Ahab does haunt the whole three-year adventure of the Pequod—but no more so than Ishmael and the whale. In my first chapter, "Loomings," you will read, "two and two there floated into my inmost soul, endless processions of the whale, and, midmost of them all, one grand hooded phantom, like a snow hill in the air." Moby Dick is "midmost, but only within an "endless procession of the whale." Indeed, even Ahab allows himself to be recalled to the wonder of the whale. In chapter 70, "The Sphynx," Ahab soliloquizes over the severed head of a whale: "Speak, thou vast and venerable head." And then, Ahab almost begins to let go of his monomania. A ship is cited and the mast head shouts, "Sail ho!" Ahab thinks to himself, "That lively cry upon this deadly calm might almost convert a better man."

The Crotch

My readers, I hope I will have them and I wish them to be with me, as it were, as I am writing my tale. Everything is important and so it is with life. Aboard an island of a ship, every piece has its place, perhaps more so than at home. I go out of my way to describe many of these details, such as the Crotch that I give in chapter 63. Imagine yourself on one of the small boats that chase the whales. This crotch supports the wooden handle of two harpoons, while the sharp ends project from the bow. The harpoons are connected by a single line. If the harpooner cannot fling both into a whale, which is usually very difficult, then the second harpoon must be tossed overboard, otherwise it will be swinging dangerously about the boat. But, there are usually three boats in the chase, each with harpoons tossed overboard. Danger is thus part of the normal routine of chasing whales.

The rhythm of these chapters, and, indeed, of much of my book, is to give us a description of either a whaling scene or of the whale itself in a particular chapter, and then, in following chapters, back up to consider details that were passed over in the first description. Thus, in the chase after a whale and in the actual killing of it, the cramped and dangerous conditions of the small whaling boats are first described in chapter 60, "The Line." Later, when we have the whole scene in mind, we proceed to details that would have slowed the adventure if they had been inserted earlier, and this occurs, for example, in chapters 61, "Stubb Kills a Whale," 62, "The Dart," and 63, "The Crotch."

Stubb's Supper

"A steak, a steak, ere I sleep! You, Daggoo! overboard you go, and cut me one from his small!" (chapter 64, "Stubbs Supper"). Yes, my carefree Stubb wants his supper from the very whale he helped catch. Of the four boats chasing this particular whale, Ahab's, Starbucks, Stubb's, and Flask's, it is Stubb's that manages the kill. His harpooner, Tashtego, stands and successfully hurls his harpoon, and then Stubb, with his dart, finds the whale's heart, and kills the leviathan. But there are dangers within and without, the rope from the harpoon can cut through one's hand, and the whale can capsize the boot.

After the killing, Stubb eats the flesh of the whale, using the light from the whale's oil. Is this a terrible thing? "Go to the meat-market of a Saturday night and see the crowds of live bipeds staring up at the long rows of dead quadrupeds" (chapter 65, "The Whale as a Dish"). So, we should all be vegetarians? Perhaps, but that is an expensive diet, and I do not think it is what we are being told. Rather, the issue concerns our hypocrisy and the ease to which we condemn others for similar deeds that we ourselves do: "I tell you it will be more tolerable for the Fejee that salted down a missionary in his cellar against famine … I say, in the day of judgment, than for thee, civilized and enlightened gourmand, who nailest geese to the ground and feastest on their bloated livers." Then too, given the needs of our time, the whale is a gigantic floating oil well: "Reckoning ten barrels to a ton, you have ten tons for the net weight of only three quarters of the stuff of the whale's skin" (chapter 68, "The Blanket").

The whale is brought alongside the main ship, saving it from sharks (chapter 66, "The Shark Massacre"); then the blubber is removed to get the oil (chapters 67, "Cutting In," and "68, "The Blanket"); and, finally, the whale is let free of the ship (chapter 69, "The Funeral"). In all this, we are brought close to the dangers of hunting the whale and to the greatness of the whale itself. Thus, my *Moby-Dick; or, The Whale* does concern the clash between good and evil, both in our own lives and in the social order? But, it does this only as part of a great adventure tale. Indeed, if my book has a single subject, it is the strange and unique embodiment of truth in a whale: "For unless you own the whale, you are but a provincial and sentimentalist in Truth" (chapter 76, "The Battering-Ram"). The whale travels the oceans and is wise in its ways.

Gams

"But look at the godly, honest, unostentatious, hospitable, sociable, free-and-easy whaler! What does the whaler do when she meets another whaler in any sort of decent weather? She has a *Gam*" (chapter 53, "The Gam"). My Ishmael then goes on to define a gam, noting that we will not find the definition in our dictionaries, although, indeed, I now observe that it is, indeed, defined in Webster as a friendly meeting between whaling ships. Ishmael's own definition, however, is more specific, namely, a meeting between two whaling ships where the captains stay aboard one ship, while the first mates stay aboard the other.

The nine gams discussed in my *Moby*-Dick are:

1. Chapter 52, "The Pequod Meets the Albatross."
2. Chapter 54, "The Town-Ho's Story"
3. Chapter 71, "The Pequod Meets the Jeroboam—Her Story."
4. Chapter 81, "The Pequod Meets the Virgin"
5. Chapter 91, "The Pequod Meets the Rose-bud"
6. Chapter 100, "The Pequod of Nantucket Meets the Samuel Enderby, of London"
7. Chapter 115, "The Pequod Meets the Bachelor"
8. Chapter 128, "The Pequod Meets the Rachel"
9. Chapter, 131 "The Pequod Meets the Delight"

Only one gam is in the group of chapters that I am considering here. Nevertheless, although some have already been mentioned and will be mentioned again, I will touch upon them all at this time. There is, however, one interesting point that should be mentioned about the title of all the gams, namely, how well they fit what happens in the gam itself. The German boat has no oil, and it is named, "Jungfrau," the Virgin; the French ship is called the "Rose-bud," and it is indeed about the smell from a decaying whale. But, then Ishmael is writing this tale, many years after the events; he has already given us a taste of his humor, and it will appear again, particularly in the strange gam relating to the Town-Ho.

Of these nine gams, seven relate to Ahab as Ahab. This is not surprising, since one of the purposes of these gams is to illustrate that captains of whaling ships have their own personal characteristics, but none approach Ahab's monomania, which evil is at odds with the very nature of a captain of a whaling ship. This becomes most evident in the gam already discussed above, namely, in chapter 128, "The Pequod Meets the Rachel," where Ahab refuses to help the captain of the Rachel look for his lost son.

That the majority of the gams relate to Ahab as Ahab might seem to detract from my claim that Ahab's monomania is not the main focus of my book. Nevertheless, they comprise only 9 chapters in a 135-chapter book, and, of those nine, only seven pertain to Ahab as Ahab. The two that do not are numbers 4 and 5 listed above. In these two, Ahab, of course, remains captain of the Pequod, and his presence as a unique captain is somewhat felt; but, for the most part, once he learns that the captains of the other ships have no information about the white whale, he leaves the running of the gams to his mates. This might be considered somewhat discourteous; but, no doubt, with all the general friendliness attributed to whalers by Ishmael, some captains might be more retired than others.

Also, there are particular reasons in these two gams why Ahab might not be expected to participate. In chapter 81, "The Pequod Meets the Virgin," a Dutch-German ship, the Jungfrau, has caught not a single whale and is actually out of oil. Its captain, Derick De Deer, carrying an empty oil can, approaches the Pequod, to have his can filled with oil. Ahab asks about his sighting of Moby Dick, but the captain has no knowledge, and prepares to depart with his can full of oil. As Captain Derick is leaving the Pequod, a whale is sighted, and he then, throws the can overboard and heads to capture the whale. The Pequod has also sighted the same whale, and it begins the pursuit. Although Derick's boat has the lead, the three American harpooners, Queequeg, Tashtego, and Daggoo, simultaneously throw their harpoons above the heads of the crew from the Jungfrau and catch the whale.

In this relatively long chapter, I once again take the occasion to discuss again the greatness of the whale that is being caught—the immense rivers of blood that pour from it, until its heart is burst—and the danger that the whale will then sink, as happens here, rather than float as is usual. Effort, pain, and no reward; but that is life. In the meantime, the Germans think they have sighted a new group of sperm whales; but, although their spout resembles that of sperm whales, they are the ultra-fast, impossible-to-catch, whales, Fin-Backs. I conclude the chapter, "Oh! many are the Fin-Backs, and many are the Dericks, my friend."

In chapter 91, "The Pequod Meets the Rose-bud," the gam is now with a French whaler, and with a stubborn and stupid captain, who insists that his crew attempt to get oil from a decaying whale, while the stench almost kills all. Ahab, having been informed that they have not sighted the white whale, leaves Stubb on board. There is a humorous scene in which Stubb and one of the mates of the Rose-bud, who speaks English, agree to deceive the French Captain about the dangers of attempting to get oil from a decaying whale. The Captain is convinced, and Stubb begins to depart, when he realizes that he can obtain some "Ambergris," the valuable source of perfume from the decaying whale. He does just that and is almost finished when Ahab tells him to hurry. But, then, any captain might have said the same, if he thought his mate was dallying and speaking to another mate.

The Town-Ho's Story

"For humor's sake, I shall preserve the style in which I once narrated it at Lima" (chapter 54, The Town-Ho's Story). Of all the gams described in my *Moby-Dick*, I grant that the strangest is the one relating to the Town-Ho, which is first introduced in chapter 54 and then mentioned again in chapter 71, "The Pequod Meets the Jeroboam—Her Story." The first mention of the gam in chapter 54 is the longest chapter in my book, and it was the only part of *Moby-Dick* that was published in magazine prior to the book itself (*Harper's New Monthly Magazine* in October 1851). But, with this comment, I must be careful not to go outside my own directive to myself of reading my great book as if I found it, lying on a beach, fortuitously near a comfortable beach-chair; with no prior knowledge about it, attracted, let us say, by the picture of a whale on the cover and by the title *Moby-Dick*.

Chapter 54 is introduced with a parenthetical sub-heading, "As told at the Golden Inn." It is a self-contained episode; but since many of the chapters have this same quality, only the length and detail might cause a reader to wonder about its general place in the story being unfolded. Still, it reads fast, and it is very strangely related to Ahab and Moby Dick, and thus above I granted the entire chapter to Ahab as Ahab. The narration of the gam consists of two parts. One part is an obvious story relating how the whaling ship, "Town-Ho," had sighted Moby Dick. The other part includes a "secret tale."

The "secret" tale centers about two sailors, Radney a mate, handsome but somewhat ruthless, and Steelkilt, a powerful but ugly sailor. Hatred has built up between the two, but each had kept their feelings in place. The boat has a leak, and Steelkilt is in charge of the men chosen for strength to handle the pumps. After his exhausting time at the pumps, Radney orders Steelkilt to clean the deck, although it is clearly not his function, but that of others who have not worked at the pumps. Radney repeats his demand, but now he is also threatening to hit Steelkilt with a "cooper's club hammer" (a hammer used to repair wooden casks), if he doesn't obey. Steelkilt does his best to dissuade the mate, Radney, but to no avail. Steelkilt then says that if the hammer touches his face, he will kill the mate. But the stupid Radney does hit Steelkilt, and the powerful Steelkilt hits Radney in the jaw—"he fell on the hatch spouting blood like a whale."

Tension—indeed mutiny—arises between those who have worked on the pumps and the rest of the crew; but the captain demands that the sailors return to work. Steelkilt, who is now the leader of the dissidents, agrees but demands that they will not be punished. Steelkilt, is, however, betrayed by his own men and beaten, almost to death, but then, whispering something to the captain, the latter stops his punishment. In the meantime, Radney sufficiently recovered from the powerful hit in the jaw by Steelkilt, continues beating Radney, even though Radney swears to kill him if he continues.

Radney plans his revenge, and he is about to execute it, when the great white whale is sighted and it is recognized as Moby Dick, whose fame has spread the oceans. All boats take part in the chase and hard feelings seem to be forgotten, even though Steelkilt is rowing the boat in which Radney, the mate is in charge. In the chase, Moby Dick causes Radney's boat to almost overturn—Radney, in fact, is tossed overboard, although the rest of the crew manages to keep the boat afloat. As Radney attempts to swim back to the boat, Moby-Dick catches him in his jaw, swims away from his pursuers holding Radney in his mouth, then dives and rises, "with some tatters of Radney's red woolen shirt."

The "presentment" of what may happen to the Pequod is clear but allow me first to jump ahead and consider the second gam, chapter 71, "The Jeroboam's Story," indirectly relating to the Town-Ho, and also concerned with the death of a mate by Moby Dick.

Because of an epidemic, Captain Mayhew, of the Jeroboam, refuses to board the Pequod, but manages to keep his small boat alongside, continuing in a conversation with Captain Ahab. Stubb, however, spots a very strange-looking person rowing the captain's boat, and he recognizes that it is the self-proclaimed prophet mentioned by those who had related to him the Town-Ho's tale. Calling himself Gabriel, he was a member of an extreme Shaker sect; he had deceived the crew of the Town-Ho by seeming to descend from heaven in their midst, but he was actually using trapdoor. Captain Mayhew was opposed to Gabriel, but he was fearful of mutiny if he dropped him off on some island.

When he was aboard the Jeroboam, Gabriel warns the crew that the white whale is the embodiment of the Shaker God and is not to be hunted. But then, Moby Dick is sighted, and, opposing Gabriel but not Captain Mayhew, the chief mate, Macey, takes a single boat after the great whale.

When Macey is standing in the bow of his boat ready to strike the whale with a lance, Moby Dick with a powerful swipe of his tale sends the mate to his doom, while leaving the rest of the boat whole and the men safe. From the ship, Gabriel proclaims the truth of his prophesy and calls the boat back.

Now, alongside the Pequod, Captain Mayhew asks Ahab if he is also going to search for Moby Dick, and Ahab replies, "Straightaway," and Gabriel warns Ahab of the "blasphemer's end." Mystery is increased when Ahab recalls that he has a letter for the Jeroboam, which turns out to be for the chief mate, Macey, who was killed by Moby Dick. When Ahab tries to give the letter to Captain Mayhew, Gabriel tells him to keep it and to give to the dead Macey himself.

What are you to make, dear reader, of these two gams, laced as they are with portents and prophecies that I will allow to be fulfilled to a great extent by the Pequod's search for Moby Dick in my attempt to please Peleg and Bildad who paid me for my writing? Let us recall again, the constant presence of these "portents," from chapter 19, "The Prophet," in which we meet Elijah, through Fedallah, introduced in chapter 50, through Gabriel, here in chapter 71, to Starbuck and Ahab himself, until the end, chapter 135.

I have already hinted at a partial "answer," namely, Ishmael is writing this narrative long after it has been written; and there is need to explain the weakness of the crew of the Pequod to stand up to Ahab. Then too, underneath this mystic, there is usually deception. Most importantly, Ahab sneaks Fedallah aboard the ship, without the owner's approval, and the strange Gabriel used deception to appear to come from heaven.

Nevertheless, these "presentments" are multilayered. First, humor pervades, almost as if Ishmael is telling us that sailors are a superstitious group, enticed perhaps by the vast oceans and sky, and the lack of land to root one's feet upon. The Town-Ho Story is introduced through an elaborate chatty atmosphere at Lima with young Don's, almost as if Ishmael is warning us not to take everything so seriously, even though it concerns a mysteriously great white whale and the deaths of those who try to kill this Moby Dick. These Spanish friends ask him to swear on the bible that his strange story is true, and he does; but, in circumstances that allow for doubt of that truth. And then, perhaps, we are meant to sympathize with the inability of the mates of the Pequod to revolt against Ahab.

Finally, amidst all warnings and deaths, we are constantly informed that they are normal in this whaling business and that any prophesy might seem to foretell what frequently happens anyway. And that leaves us exactly where, I think one should be in relation to my book, namely, drawn into a tale about whales, about leviathans, whose size would overshadow the great dinosaurs, if they were still alive, and whose speed through the oceans of the world might embarrass the flights of the greatest birds. Can mystery be far from these giants and can an almost religious atmosphere not surround their pursuit by man?

CHAPTER 10

Whales! Conversation, Art, Dining, Business, and Poetry

Abstract The critics who maintain that Melville's book, *Moby-Dick; or, The Whale* is mainly about Ahab's crazy pursuit after the white sperm whale, Moby Dick, assert that these and similar chapters are intended to prepare us for believing in the power of Moby Dick to destroy a ship. This assertion is nonsense, and it will become more and more nonsensical as we proceed. Such a detailed attention to whales and whaling in general cannot be justified as merely preparing for us to believe in the power of sperm whales to destroy a ship, when, in fact, this truth has already been given to us from the "Extracts."

Keywords Ahab • White Sperm Whale • Moby Dick • Loose Fish

> One often hears of writers that rise and swell with their subject, though it may seem but an ordinary one. How, then, with me, writing of this Leviathan? Give me a condor's quill! Give me Vesuvius' crater for an inkstand! Friends, hold my arms. (chapter 104, "The Fossil Whale")

(Chapters 72–105 of my Moby-Dick)

J. S. Catalano, *In Pursuit of Moby-Dick*,
https://doi.org/10.1007/978-3-031-40357-6_10

Yes, dear reader, coming after 103 chapters and two preliminary sections, it is you who might need help in holding what has been written. But we have a big book for a big subject. The leviathan swims and owns the oceans of the world, and we follow this leviathan with our hearts, our books, and our harpoons. Should we leave our harpoons home? Perhaps, we should. But then let us stop defiling the Earth in search of oil. Moreover, in their capture, every part of the whale is used; and the whaling men pit their lives in their hunt. "The more I dive into this matter of whaling," writes Ishmael in chapter 82, "The Honor and Glory of Whaling," "so much the more am I impressed with its great honorableness and antiquity."

And, indeed, I, Melville, and my spokesman, Ishmael, dive into this matter of whaling. I do it with such exuberance—with such careful and consistent attention to all the mates, harpooners, and sailors of the Pequod, as well as to the details and comparison of sperm and right whales they hunt—that the total ensemble of whales and American Whaling must find its equal place in this great narrative alongside Ahab's mad search after Moby Dick. I have planned it so that chapter after chapter, dramatic episode after dramatic episode, effectively draws us away from Ahab's lunacy into the specific dangers and bravery of the men of the Pequod. And to this, one must add my almost mystical belief in the whale. Yes, I keep informing my readers of something that I am afraid they will not believe—*whales exist!* Our mythology, our Bible, our literature, each proclaims the majesty of whales and the wonder of our encounter with them.

Thus, do not let your mind race ahead in the knowledge that I let the Pequod be destroyed by Ahab's monomaniacal pursuit after the great white sperm whale, Moby Dick, as, to repeat, I write an ending to my tale to satisfy those paying me to write, Captains Peleg and Bildad—and to a hidden ending more to my liking in which the Pequod returns. Yes, *in my writing of the book, Ishmael will let the Pequod will be destroyed, and yes, that is important.* Nevertheless, during the long voyage, honorable deeds will be performed, and that last great tragedy does not diminish their value anymore than a bomb which destroys a neighborhood diminishes the good deeds done earlier that very day.

Sperm

For example, in chapter 78, "Cisterns and Buckets," we are given details of a rescue by Queequeg of the Indian harpooner, Tashtego, and then the rescue of both by the giant black harpooner, Daggoo. The dramatic scene follows the strangely titled chapter "The Great Heidelburg Tun," and, although I do not give a note explaining the term, the context of the chapter gives us a fairly clear meaning of its use, namely, a gigantic and beautiful barrel that held the best wines of the "Rhenish valleys." The comparison is with this barrel and the forehead of the sperm whale, with the vast quantity of sperm it holds. (This term, "sperm" becomes so common in my narrative, that its sexual meaning is almost lost in favor of its human uses.) In his attempt to cut into the whale's head, moored as it is alongside the Pequod, Tashtego falls into the cavity holding the sperm. Queequeg immediately jumps in to rescue him, but he has to cut his way through with his sword to find Tashtego's body. In the meantime, Daggoo, risking his own life, assumes a precarious position to retrieve both as soon as he can spot their bodies. Luckily, Daggoo is successful, and he lifts them both on deck.

Reflections and humor are intertwined in a special way in this chapter, 78. There is a comparison of "midwifery" with fencing, since Queequeg had to cut his way through the sperm to rescue Tashtego. The chapter ends with a more philosophical reflection, "How many, think ye, have likewise fallen into Plato's honey head, and sweetly perished there?" Plato suggests that our earthly life merely reflects a deeper drama of an earlier life—a life in which we were banished from a spiritual realm because of some great fault, and must now find our way back. But, the real substance of the chapter is the drama and the bravery of the rescue, on the one hand, and, on the other hand, the immensity of the sperm whale's sperm—"five hundred gallons."

Owning the Whale

> Now, mark ... when I shall hereafter detail to you all the specialties and concentrations of potency...when I shall show you some of his more inconsiderable braining feats; I trust you will have renounced all ignorant incredulity ... For unless you own the whale, you are but a provincial and sentimentalist in Truth. (chapter 76, "The Battering-Ram")

Consider again the headings of many of these chapters; they are all about whales and whaling: to give a few, "The Sperm Whales Head," "The Right Whales Head," "The Honor and Glory of Whaling," "Jonah Historically Regarded," "The Tail," and "Stowing Down & Cleaning Up." I hope that you are not particularly interested in what might have spurred me to write this or that chapter. Rather, I desire that my readers will note that from chapter 72 to chapter 105, there are thirty-three chapters of which *thirty* deal exclusively with whales and whaling (I grant chapters 73, 99, and 100 to Ahab as Ahab).

If there is a key to this group of chapters on whaling, I think that it is given in chapter 76, "The Battering-Ram," which is a description of the power of the sperm whale's head, and in which I write, "For unless you own the whale, you are but a provincial and sentimentalist in Truth." What do I mean by this claim, which I have given several times above? Granting the fruitful ambiguity that can arise from literature, I want my readers to be reminded of both the obvious and hidden existence of whales. All things are both obvious and hidden. A snow flake falls; but who can describe its uniqueness, who can understand its mystery? Even if the weather conditions are right and we can discern an individual snow flake, we still view it from only one perspective. But this reflection, while true and important in other contexts, is too "sentimental" for our tale. The pursuit of these giant sperm whales is, after all, a life and death contest that reflects our individual and social lives.

Again, I am not here concerned with interpretation; I do not go outside my own text, and I do not think this necessary for my readers. I am here in this book a generous writer; hiding nothing from my readers except the ending of my tale, and there I give generous hints of another ending. Yes, dear reader, be patient and awake, first, to the brute existence of things—the way whales simply live their greatness unaware of our existence; second, to our vision of this greatness—for we are the eyes of the universe; and third, to our need to understand, to classify, and to use these leviathans—for we are the mind, the voice, the pen, and the hands of the universe. And, thus, it is that we question what we see when a sailor shouts, "There she blows!"

What is it that blows? I inform my readers that no concrete experimentation can solve the mystery of the whale's spouting. Thus, we have a whole chapter, 85, "The Fountain": "My dear Sir, in this world it is not so easy to settle these plain things. I have found these plain things the knottiest of all." We then have a few, rather heavy pages devoted to the nature of the whale's spouting, pages that need a big dose of humor We should not be surprised at the whale's spouting, for all great thinkers, including Ishmael, spout. "I had the curiosity to place a mirror before me; and ere long saw reflected there, a curious involved worming and undulation in the atmosphere above my head." And, you dear reader, if you sometimes abandon yourself to deep thoughts about life and its meanings, would see the same mist, although it may take the form of a smile or a frown. These mists are like the doubts we may have about the meaning of life and the truths we sometimes suspect exist in these doubts:

> Doubts of all things earthly, and intuitions of some things heavenly; this combination makes neither a believer nor infidel, but makes a man who regards them both with equal eye.

It is, however, important not to be so captivated by the metaphysical reflections, which occur throughout my narrative, that one loses sight of what they are based—the truth of whaling. Indeed, both are important; reflections are the soul of this narrative and whaling is its body. But the union of body and soul are one; every sailor's gesture and every movement of the whale unite in the same adventure, *Moby-Dick; or, The Whale.*

A Loose Fish?

"What are the Rights of Man and the Liberties of the World but Loose-Fish? ... And what are you, reader, but a Loose Fish and a Fast Fish, too?" (chapter 89 "Fast-Fish and Loose-Fish"). After an entire chapter praising the majesty, beauty, and power of a sperm whales tail—"The more I consider this mighty tail, the more do I deplore my inability to express it" (chapter 86, "The Tail")—and after describing the behavior of large groups of whales in chapters 87 and 88, "The Grand Armada" and "Schools and Schoolmasters," I devote two chapters, 89 and 90, to the legal question of who owns a whale that is pursued, and harpooned, and yet escapes one ship, only to be finally caught by another ship. As we might by now expect, this legal question concerns more than whaling, although the reflection does indeed arise from whaling.

Chapter 89, "Fast-Fish and Loose-Fish," gives us two laws, both of which seem to fit commonsense, namely, "A Fast-Fish belongs to the party fast to it, and a Loose-Fish is fair game for anybody who can soonest catch it." Ishmael grants that, although brief, both laws need explanation. In the final analysis, a whale belongs to the ship that demonstrates its intention and ability to make a whale "fast" to the hull of the ship, that is, to bind it to the ship. And, everything within, tied to, or, in any way connected to the whale—harpoons and other whaling equipment—belongs to the same ship. Thus, if a ship harpoons a whale, which then escapes, the whale and the harpoons belong to the ship that finally catches the whale. I illustrate the two laws with relation to the laws of marriage and divorce, and to other legal matters; but I grant that I may have let myself here go beyond what is called for and what can be justly affirmed as truth.

On the other hand, I so develop the chapter that it pertains more fruitfully to the relation of this whaling law to all laws binding human behavior; namely, whatever is said to the contrary, possession is not only half the law but too often the whole law. And this applies not only to the hard earned "widow's last mite" but to whole continents: "What was America in 1492 but a Loose-Fish?"

You must allow me, dear reader, to go on with a subject once I get hold of it. You may say to yourself, yes, but you made this point and why repeat it so often. I repeat the truths about whales because they are unbelievable to any normal logic; they should not exist—only they do, and they lead naturally to us thinking about ourselves. With this in mind, let me approach this chapter that I have written in stages. Thus, before I give a brief summary of the chapter, I wish to say something about the atmosphere in which it is written. It reports to be a reflection upon an English law concerning who owns a whale on the shores of any English territory. I say "reports to be," because the text itself warns us when the reflections on whales and whaling may be stretched outside the realm of the truth about whales, and this is one of those times.

This distinction between the real facts about whaling and possible exaggerations is important, and although I have made it before, I wish to emphasize it again here. For example, the previous chapters on the spout and the tail, the whales traveling in groups, indeed, most of the chapters devoted to whales and whaling in general, are based upon truth. And, the unique force of *Moby-Dick; or, The Whale* arises from that truth. Whales are the leviathans of the world, and the American whalers of the nineteenth century were engaged in a glorious hunt, in which they respected their prey. True, there is a concern for money, but once again let us put this in proper perspective before we proceed.

Whatever one wishes to say about the magnetic draw to profit, it is not the issue in this tale. Each sailor will profit from the voyage, and, if the "lays" are not always as generous as they should be, at least Captain Peleg was decently generous with Ishmael and very generous with Queequeg, "we'll give ye the ninetieth lay, and that's more than ever was given a harpooner yet out of Nantucket" (chapter, 18, "His Mark"). No, as already mentioned, the killing of whales is not to be compared, for example, to the senseless killing of buffalo.

With all these caveats in mind, let us now return to the context of Chapter 90, and why one might consider it strange. It is sort of a corollary on the previous chapter concerning who owns a whale that has been searched by one party but caught by another. The specific allusion is to an English law claiming that any whale on English territory, for any reason whatsoever, belongs to the King and Queen, as follows—the head to the King, the tail to the Queen. I note that, as a matter of fact, this completely divides the whale, so that there is nothing left for those who might have spent time and effort to bring the whale on shore.

What, for an astute reader is a warning that this reflection may or may not be based upon fact is the Latin quote heading the chapter, which is explained generally in the first paragraph. But this should warn my readers that, whether founded on fact or fiction, I have another motive in mind. Again, this is not an issue of interpretation but of reading the text a little more carefully. This strange English law about whaling, which led me to give us a separate chapter, is on the same level that prompts the English railways to always keep a separate car for the possible use of royalty. That is to say, the entire chapter harpoons British snobbery in general, rises again, and then harpoons British royalty, the King in particular. For the law notes that it is not only the whale that belongs to the king and queen but the sturgeon—the king should be given its head, "because of the highly dense and elastic head peculiar to that fish.... And thus there seems a reason in all things, even law," and, even in this chapter.

The Castaway

> Already has it been related how the great leviathan is afar off descried … how, in due time, he is condemned to the pots … but now it remains to conclude the last chapter of this part … singing if I may. (chapter 98, "Stowing Down and Cleaning Up")

Chapter 98, "Stowing Down and Cleaning Up," is the last chapter of "this part"; but where does the part begin? Since this is not a textbook, and since the whaler me, Melville, as Ishmael, the metaphysician from whose pen the reflective Ishmael guides our story, we should not be surprised that there is not one beginning but two or three. Two killings of sperm whales by the crew of the Pequod are described in the preceding chapters in great detail. The nearest one to this chapter (98) begins in chapter 93, "The Castaway." But the reference to concluding "this part," may refer to its beginning in chapter 60, "With reference to the whaling scene shortly to be described," which, in fact, refers to chapter 61, "Stubb Kills a Whale," and continues through chapters 70, 71, and 73.

Ahab has, of course, been present throughout as the captain of the Pequod, and his mad search after Moby Dick has not been completely forgotten. Nor do we lose track of the tragic ending about to overtake the Pequod. Nevertheless, this "interruption" is too long to be given to us as a pause to catch our breath before we go follow Ahab to his doom. And, the details are too inherently interesting to be regarded as a mere side-show to Ahab as the main event.

Thus, chapter 93, "The Castaway," describes a particular hunt after a sperm whale; it is an irreducible drama in its own right, although from another perspective it can be seen as anticipating the last tragic drama of the Pequod. A sperm whale has been sighted, and we are informed about a division within the crew of a whaling ship, which has never been explicitly explained before, namely, those who are in the small boats that chase the whales, and those who stay on board the main ship—"the ship-keepers." In general, these later are as important as those who man the small boats, for if the ship is lost, all are lost.

We are informed that if any person should be fragile, he will be a ship-keeper, and this was the role of the cabin-boy, Pip. But, Stubb's normal "after-oarsman" had sprained his arm, and Pip was chosen to take his place. For a while, the frightened boy avoided danger; but then, in the heat and danger of the chase, he jumped overboard and became tangled in the harpoon line, which had to be cut to save his life. The whale was lost.

Stubb then told Pip that if he jumped again, he would be left. And, indeed, Pip did jump again. He was not tangled in any line, but was merely left as a "castaway." Nevertheless, we are informed that it is common in the chase that a crew member will be left behind because there are usually other boats chasing the whale and one of these can bring the lost one on board. Or, the ship can do it, if one can swim to it. It was the ship that rescued Pip, but not before he almost drowned, thereby losing his reason, or, at least, so it seemed to all. But, my Ishmael uses the occasion of this near drowning to give us a reflection on wisdom, sanity, and insanity: "He saw God's foot upon the treadle of the loom, and spoke it; and thereby his shipmates called him mad." Yet, to repeat, in a strange way, the mad cabin-boy, Pip, then becomes close to the mad Captain, Ahab.

Ishmael brings us alongside him as he is experiencing his voyage on the Pequod; but he also continually whispers in our ears that, yes, all this happened sometime ago, and, perhaps, we were not all the men we should have been, or, the final tragedy might have been averted. On the other hand, whaling is dangerous, there are deaths, and sperm whales do stow in ships. What happens in the normal course of this dangerous business of whaling is one thing, but to allow one man's monomania to finally rule the day is another thing. And yet all the evil power in Ahab's soul could not diminish "The Honor and Glory of Whaling" (chapter 82).

A Movement of My Twice-Told Tale

"The skeleton dimensions I shall now proceed to set down are copies verbatim from my right arm, where I had them tattooed" (chapter 102, "A Bower in the Arsacides"). I have mentioned this chapter a number of times, and I think that, with the first chapter "Loomings," it gives one of the keys to a possible multileveled reading to my tale—a term which I hesitate to use in regard to my book that is, I believe, uniquely lucid and generous in what it offers us on surface of the text. But, allow me to repeat some of the strange aspects of this rather long chapter—indeed, it is all strange, and meant to be that way. Yes, I repeat, but let it stand. Ishmael reports that, after his adventure with the Pequod, he had many wild wandering, and in one of them he met "a royal friend Tranquo, king of Tranque, one of the Arascides." We are then told of a temple built from the skeleton of a stranded sperm whale. Ishmael enters and begins to measure its length: "The skeleton dimensions I shall now proceed to set down are copied verbatim from my right arm, where I had them tattooed; as in my wild wandering at that period, there was no other secure way of preserving such valuable statistics."

On the other hand, the very next chapter, 103, "Measurement of the Whales Skeleton," has nothing to do with those tattooed measurements, and indeed, we have been given accurate first-hand accounts of the size of the sperm whale much earlier, for example, in chapters 61, 68, and 74, and, indeed, almost throughout the voyage. Thus, what precisely is the point of chapter 102?

Simply put, are we to actually believe that Ishmael tattooed the dimensions of a sperm whale on his right arm? Would this not put a spurious light on all the facts about sperm whales and American whalers, facts that Ishmael goes out of his way to inform us repeatedly are not exaggerations, but the literal truth. In some way, the unity of my book requires that these exaggerated claims, with their bizarre humor, stand side by side with all the carefully detailed truths about whales and whaling. For example, recall that doubloon, nailed to the main mast by Ahab (chapter 99), a doubloon true in its gold, but with astrological figures. It too must have its own particular history of fact and fiction: “It so chanced that the doubloon of the Pequod was a most wealthy example of these things. On its round border it bore the letters, REPUBLICA DEL ECUADOR: QUITO” (chapter 99, “The Doubloon”). Chapter 99 gives us reflections upon the doubloon, which, in chapter 36, “The Quarter Deck,” Ahab nailed to the mast head. I will give only part of Ahab’s own reflections: “This coin speaks wisely, mildly, truly, but still sadly to me. I will quit it, lest Truth shake me falsely.” And so Ahab again flees self-knowledge.

And, to jump ahead, in chapter 115, “The Pequod Meets the Bachelor,” in which we are given a brief insight into another whaling voyage—this one happy and with dancing Polynesian girls who had eloped with the sailors—we read this comparison of two captains:

> Lord and master over all this scene, the captain stood erect on the ship’s elevated quarter-deck, so that the rejoicing drama was full before him, and seemed merely contrived for his own individual diversion.
>
> And Ahab, he too was standing on his quarter deck, shaggy and black, with a stubborn gloom.

“Stubborn gloom” is a well-chosen phrase, hinting as it does of Ahab’s constant renewed vow to remain true to his monomania.

But, what, my patient reader, is the place of fiction in life? Our cosmopolitan life, our cities and our ships, our churches, and our new cathedrals—banks, banks, and more banks—are built as much by beliefs as by the truths of science. Thus, throughout my tale, we are given instances of the ways architecture embodies our beliefs—whether it be features of the church, as described in chapters 7 and 8, “The Chapel” and “The Pulpit,” or of the whaling ship, as described particularly in chapter 96, “The Try-Works.”

CHAPTER 11

Ahab's Leg and Ahab's Life

Abstract These last three chapters, "The Chase: First Day," "The Chase: Second Day," and "The Chase: Third Day," have managed to swallow the previous 132 chapters not only in the minds of the public who have not read the book but even with too many critics. If Ahab's monomania led to the destruction of all the crew of the Pequod except Ishmael, then a monomaniacal reading destroys the very life of the book. Nevertheless, they are the three concluding chapters, and they deserve proper attention. Indeed, they are a fast-paced read!

Keywords Ahab • Chase • Parsee • Prophet • Job • Carpenter • Blacksmith

> Tashtego, Queequeg, Daggoo! What say ye, pagans! Will ye give me as much blood as will cover this barb? (chapter 113, "The Forge")

At the end of the gam with the Samuel Enderby and while leaving the British ship to go back to the Pequod, Ahab damaged his whale-bone leg. Now on board the Pequod, he calls upon the carpenter to mend it, and

(Chapters 106–135 of my Moby-Dick)

J. S. Catalano, *In Pursuit of Moby-Dick*,
https://doi.org/10.1007/978-3-031-40357-6_11

the carpenter calls upon the blacksmith to do the "buckle-screw." You would not be wrong in concluding that the two scenes are like a Greek-inspired drama. In chapter 107, the carpenter is introduced to us as someone who has made himself into a human-tool, to be used as needed by all on board the Pequod. Yet, in the last paragraph of that chapter, I remind all that, nevertheless, he was "no mere machine of an automaton."

There then follows the dramatic form of chapter 108, "Ahab and the Carpenter," in which the down-to-earth carpenter stands in stark contrast to the reflective Ahab. Ahab asks the carpenter to explain how one can feel a leg that is missing, and the carpenter simply responds that it really is a puzzle. Ahab then goes on to speculate that the carpenter himself may be only the shadow of another and greater thinking thing; and, further, perhaps, after all, one may feel pain in hell without a body. To which the carpenter replies that he must have miscalculated the measurement of Ahab's leg.

Perth

The carpenter is simply "the carpenter"; I do not give him a personal name. In contrast, the blacksmith has a name, "Perth," and he is, like Ahab, both reflective and a tragic figure. He is formally introduced to us in chapter 112, "The Blacksmith." He was a well-respected, happily married, and prosperous until he was robbed of all his possessions by someone close to him. He then lived on to see his wife and children die in poverty. Perth then takes to the sea and works at what he is best at, being a carpenter.

Ahab senses the depth of Perth's misery; and, in the following chapter, 113, "The Forge," where I have the scene continued, Ahab, seeing the sparks from the forge strike Perth, asks why they do not scorch him. Perth replies that he is scorched all over anyway. Listening to Perth, Ahab says to him, "Thy shrunk voice sounds too calmly, sanely woeful to me. In no Paradise myself, I am impatient of all misery in others that is not mad."

If Ahab is mad, and, indeed he is, it is not a madness that sweeps over him, robbing him of his reason; on the contrary, his madness is one with his cunning and his self-knowledge. That is to say, Ahab's madness is chosen and held in existence by him, as something necessary to obtain his goal. And, yet, Ahab is not so hardened in his madness that the door to true self-knowledge does not now and then begin to open. Ahab knows that Perth suffered and went to sea, if not for a cure, at least to find some joy in a life of doing what he was best at doing when he was prosperous. Ahab knows that this course is open to him—he is a good hunter of whales, despite his lame leg; but he chooses rather to nurse and expand upon his sufferings, *lest they become all-too-human.*

Still, Ahab continues to watch Perth hammer and smooth out an old "pike-head," and he asks Perth, "And I suppose thou can'st smoothe almost any seams and dents; never mind how hard the metal, blacksmith?" "Aye, sir, I think I can; all seems and dents but one," replies Perth. Ahab then, dramatically sweeping his brow, asks Perth whether he can smooth out that seam? Perth replies: "Oh! that is the one, sir! Said I not all seems and dents but one?"

I would have my readers become aware that Ahab's encounter with Perth also provides us with another insight into Ahab's self-pity and self-deception. He would have us believe that he would gladly lay down his head on the anvil and have Perth hammer away, if only he could stop thinking about Moby Dick. But, he chooses to keep thinking about Moby Dick, despite the numerous and real tasks of whaling that could absorb him, or at least distract him, if indeed he wished to be distracted. No, the truth is that he, Ahab, would be lost without his monomaniacal pursuit after Moby Dick. And the proof that his seeming self-pity is but a spur to his hate is soon to follow.

Perth finishes his task, and he is about to temper the "buckle-screw" in water. But Ahab seeks to temper it in blood, and he asks the three legitimate harpooners (not Fedallah, interestingly) to provide the needed blood, and they do so—"Three punctures were made in the heathen flesh." Then, Ahab continues with an attempt at a diabolical blessing on the harpoon, "Ego non baptizo te in nomine patris, sed in nomine dialboli." (I do not baptize you in the name of the father but in the name of the devil.) Ahab's action is dramatic, but it doesn't amount to much, a little blood voluntarily given and a few pompous words.

But, I would also point again to the consistent rhythm affecting the unfolding drama, and, indeed, my entire book. Logically, one might think that Chapter 108, in which the carpenter has been introduced and the blacksmith mentioned, would be followed by the two chapters that discuss the blacksmith, namely, 112, "The Blacksmith" and 113, "The Forge"; but two other chapters intervene—110, "Queequeg in His Coffin," and 111, "The Pacific." Furthermore, directly after the dramatic chapter, 113, "The Forge," we have the light-hearted chapter, "The Gilder:" "I am

Stubb, and Stubb has his history; but here Stubb takes oaths that he has always been jolly!"

This rhythm takes the bite out of Ahab's attempt to make himself a devil's disciple. For, indeed, the real evil is yet to happen. True, Ahab had earlier sneaked a private crew and harpooner aboard the Pequod, and that was a deceitful act. But that too might have been passed over, if Ahab had changed his mind about hunting Moby Dick. To repeat, the real evil of that hunt appears for the first time in the gam with the Rachel (chapter 128, "The Pequod Meets the Rachel"). Ahab's maniacal search for the one whale, Moby Dick, now leads to a clear act of injustice. We are no longer in the realm of grand posturing but of deed. Let us recall, that Captain Gardner has lost his son at sea, and asks Ahab for help to find him. Stubb, who first thinks Gardner had lost some small personal object like a coat or watch now says, "'His son!' cried Stubb, 'oh, it's his son he's lost! I take back the coat and watch—what says Ahab? We must save that boy.'" But Gardner has also sighted Moby Dick, and Ahab, not wishing to lose a minute in his chase, stresses his rejection of giving help, "in a voice that prolongingly molded every word—'Captain Gardner, I will not do it.'" Starbuck is witness to this scene; and, he now realizes clearer than ever before that Ahab has lost his right to rule the Pequod. Yet, he obeys Ahab's command to "let the ship sail as before."

Mutiny?

"I stand alone here upon an open sea, with two oceans and a whole continent between me and the law" (chapter 123, "The Musket"). Even before the gam with the Rachel, Ahab's sins had multiplied, and Starbuck sees himself becoming part of them. Ahab does not own the Pequod, and the owners would neither have authorized his monomaniacal pursuit nor given their approval of a special boat and crew. Starbuck knows that, however much these owners hate mutiny, they would have approved of his taking over the ship. No doubt, they could have attributed Ahab's monomania to illness, and would have absolved both Ahab and Starbuck.

Still, what a demanding deed is mutiny! As chief mate, it is he and he alone who must bear the burden of the decision. The inward and personal tension he suffers is almost unspeakable. It is the drama of each single person standing alone for truth. For, even if the law is within arm's reach, it can be oceans away in unjust prohibitions and in power plays that submerge the poor in favor of the rich. For Starbuck, the separation from the judgment of law is more concrete—oceans and continents away—and this separation weakens his self-reliance. In chapter 123, "The Musket," he enters Ahab's cabin, while the captain is asleep. He pauses before the rack with the muskets, fixing his gaze upon the one which Ahab had used to threaten him, and all the possible avenues for mutiny pass before his mind. He does not question the justice of such a move, but only the means and the fortitude needed to carry out the deed. Starbuck will not murder, for that goes beyond mutiny and it goes against Starbuck's own sense of justice.

Still, how does one seize control of a ship from such a man as Ahab? Starbuck reflects, how does one cage Ahab? "I could not endure the sight; could not possibly fly his howling; all comfort, sleep itself, inestimable reason would leave me on the long intolerable voyage." But Starbuck here allows himself to be deceived. He seems to have almost no doubt that he and the crew will be destroyed, or, at least, this seems to him to be a high probability, and yet he allows himself to be distracted by these secondary thoughts of how to put up with a caged Ahab.

Starbuck does object frequently to Ahab's behavior and to his mad pursuit after Moby Dick, but it is only with words. And, at times, Starbuck has no words to object to Ahab, only a look. In chapter 124, "The Needle," a violent storm has switched polarity of the magnetic needle so that it points east when the Pequod is going west. Ahab alone notices the error, and he is aware that the crew again needs a show of his power to keep it focused on his mad pursuit after Moby Dick. He, thus, knocks off a steel head of a lance, and proceeds to hammer it, knowing it will eventually respond to the earth's magnetic field. Starbuck, knowing that this apparent magic is nothing but good science, gives the only disproval he can muster and looks away. But the common crew stars in wonder, "one after another they slunk away. In his fiery eyes of scorn and triumph, you then saw Ahab in all his fatal pride."

"Scorn and triumph" are here as important to note as "fatal pride." Ahab goes far beyond the natural pride of being a captain of a whaling ship, who is king of the "knights and squires—the mates and their harpooners" (chapters 26 and 27, "Knights and Squires"). Ahab is becoming evil, and his main knight, Starbuck, must seize power, temporarily for himself. Mutiny and revolt, however, are easy words to pen but difficult deeds to accomplish. And yet, we witness the ashes left from Starbuck's moral weakness: "Alike, joy and sorrow, hope and fear, seemed ground to finest dust, and powdered, for the time, in the clamped mortar of Ahab's iron soul" (chapter 130, "The Hat"). Ahab has won. Starbuck and the Pequod have lost, whatever the real fate. And yet, Ahab, with all his sharp reason, is troubled by omens.

Omens, Real and Imaginary

"But I said, old man, that ere thou could die on this voyage, two hearses must verily be seen by thee on the sea; the first not made of mortal hands; and the visible wood of the last one must be grown in America" (chapter 117, "The Whale Watch").

Here then is Fedallah's prediction about Ahab's death. It is specific, and *I will make it all come true in all its details.* We can credit chance, cleverness, or the powers of a mystic; but it springs rather from something deep within the tale itself. We are thus brought once again to reflect upon the constant presence of portents in *Moby-Dick.*

Portents and omens arise in the first chapter, "Loomings," and they continue throughout the narrative (although they are mostly absent in the long reflections on whales and whaling). Yet, even where they are present, their force remains ambiguous, and deliberately so. Thus, Ishmael *enters* the Spouter Inn, despite the ominous name of its owner, "Peter Coffin." He *boards* the Pequod, despite Elijah's warnings. And here, Ahab continues to chase Moby Dick, even when he sees Fedallah's predictions about his death coming true before his very eyes. Portents and omens abound, and yet no one alters their behavior because of their presence.

Further, Ahab is aware of the foolishness of heading omens, "Omen? omen?—the dictionary! If the gods think to speak outright to man, they will honorable speak outright" (chapter 133, "The Chase—First Day"). And yet, he cannot dispel their presence from his own mind: "The things called omens! … Oh! how valently I seek to drive out of others' hearts what's clinched so fast in mine!—The Parsee—The Parsee! Gone, gone? … *I'll, I'll* solve it, though'" (chapter 134, "The Chase—Second Day").

Although Fedallah's predictions do not influence Ahab's actions, they are different from all the preceding ones in being so specific; and, in coming true. Furthermore, unlike Elijah's warnings, they have no foundation in fact—Elijah probably did see Fedallah and his crew sneak on board the Pequod. Finally, Fedallah's bond to Ahab is so strange that it may be difficult for a careful reader to accept both him and his predictions as part of my narrative. But they are part of this narrative *that I wrote, yes, for my pleasure but also to please those supporting my writing, Bildad and Peleg.*

Rightly, would one wonder if Fedallah's predictions are to be accepted as a normal part of the discourse of this tale? Or, does Fedallah's presence hint at something deeper in my tale, *another ending*?

I come back for the last time to the question of omens and portents. They occur frequently; and, at least, here, in regard to Fedallah's predictions about the Pequod, they are too precise to be accepted on face value, even as part of a normal fictional narrative. But, as part of a distinctive, twice-told narrative, they may have a special function. To repeat, other tales have been narrated by one who witnessed the events and only later puts them down on paper. Here, however, Ishmael is clearly adding information to a tale written in both the present and past tense; and he lets us know that he is doing just that, if we read carefully.

Most of what Ishmael adds is not only harmless, but easily acceptable for the tale. Yes, to repeat, I needed a seasoned narrator to give us all that information about whales and whaling together with their reflections. The Ishmael writing must have more experience than the Ishmael who first took the voyage on the Pequod. All of this is acceptable; but, then, Ishmael gives us too many disturbing additions to that first voyage. Recall some of the most obvious instances (after that given to us in the first chapter, "Loomings"). For example, chapter 54, "The Town-Ho's Story," Ishmael is very explicit about his humor: "For humor's sake, I shall preserve the style in which I once narrated it at Lima, to a lounging circle of my Spanish friends, one saint's eve, smoking upon the thick-gilt tiled piazza of the Golden Inn." And, then, at the end of this, the longest chapter in the book, the audience asks Ishmael to swear that the narrative about Radney, Steelkilt, and Moby Dick is true. Ishmael agrees to swear on a bible, and he asks that they bring "the largest sized Evangelists you can." He then swears to the truth of the tale, for he was on the very boat, and has "talked with Steelkilt since the death of Radney." But, which audience is Ishmael trying to convince? I think that it is you and I, dear reader. That is to say, he is telling us not to be convinced that any of this is part of the tale of blubber, truth, and poetry.

Most explicitly of all, how can we escape the unsettling wonder and force of chapter 102, "A Bower in the Arascides?" Ishmael knows all about whales, not only from this first voyage on The Pequod but from other later voyages, which gave him first-hand knowledge. Fine, that is acceptable! We, the reader, do not need the tale of the unbelievable encounter with a temple made from a sperm whale's skeleton. Ishmael tells us that he tattoos the measurements of the skeleton on his right arm, "as in my wild wanderings at that period, there was no other secure way of preserving such valuable statistics … I did not trouble myself with the odd inches, nor, indeed, should inches at all enter into a congenial admeasurement of the whale." That is to say, inches or yards of narrative should not enter into a tale about whales. And, yet, it is the Truth that I, Melville, aim to give my readers, and he succeeds! But how does he succeed in this twice-told tale?

There are two tales within my *Moby-Dick; or, The Whale*; and who or what is the mediator? In the tale itself, the whale is the mediator. On another level, my readers are to be the mediator. On this level, I do touch interpretation, and I will continue with this reflection in my concluding sections. But, at least here we can clarify once more the two tales themselves. One tale more or less attempts to describe the voyage Ishmael actually took on the Pequod, with later reflections that easily fit within the tale itself. Thus, in the "Extracts," we have a great deal of information about whales and whaling that a young Ishmael would not have known, and this kind of information is continued in the tale. For example, much of the information in the long series of chapters about whaling could not have been known by a sailor on his first trip—chapter 55, "Of the Monstrous Pictures of Whales"; 56, "Of Less Erroneous Pictures of Whales," and so on. These additions easily fit within the tale.

But those other, "wild" additions are on a different level—"A Bower in the Arascides?" These do not fit within the normal discourse. And, yet, both these tales are woven within the one narrative, *Moby-Dick*, and, to repeat, the mediator, in the text itself, is my reader and the whale! For, more than half the book is a straight-forward eulogy of the greatness of whales and of the majesty of the sailor's who pursue them. But that mediation still leaves the haunting presence of omens and portents somewhat unresolved. For the present, let us return to the concluding chapters of the book as written for those who paid me to write, Peleg and Bildad—their aim, if truth is to be known is to redeem Ahab who, perhaps, they should not have allowed to sail so soon after his wounded leg.

Ahab Choosing Evil

> Oh, my Captain! my Captain! noble soul! grand old heart, after all! why should any one give chase to that hated fish! Away with me! Let us fly these deadly waters! Let us home. (chapter 132, "The Symphony")!

Small good deeds are usually scattered among a life of evil, and, moreover, until death, there is always the possibility of change. Thus, as we have already seen, after the great storm, calm settles on the ocean and the air, a calm that seems to soften Ahab's soul: "Oh, Starbuck! it is a mild, mild wind, and a mild looking sky." Ahab continues talking to Starbuck, reflecting upon his life—a boy-harpooner of eighteen, who on such a day as this struck his first whale and who has not been on land for more than three years out of forty. He feels old, worn down by centuries of evil as if he were Adam cast from paradise bearing the present and future sins of the world. He asks to look into Starbuck's eyes, and he sees there a longing for home, "No, no; stay on board, on board!—lower not when I do." Then Starbuck replies as above, "Oh, my Captain! my Captain! noble soul…."

Starbuck tries once again to turn Ahab away from his purpose, but Ahab allows his mind to wander into fate and destiny. Is he, Ahab, really responsible for what he seeks, or has this mad search of his been dictated by fate from all eternity? In truth, Ahab knows that the quest is his and his alone. But, like his conjuring tricks, Ahab now attempts to fool himself. Starbuck, however, realizing that his captain's reflections are a mere passing fancy and that nothing is going to be changed, walks away, "blanched to a corpse's hue with despair." Ahab turns, sees that Starbuck has gone, begins to cross over to the other deck, but then sees the reflections of Fedallah's eyes staring back at him; for Fedallah had been leaving over the same rail. The chapter thus ends, but one cannot but wonder if Fedallah was watching to see if Ahab would indeed weaken in his purpose.

If incapable of revolt, it is to Starbuck's credit that he does not give up on attempting to alter Ahab from his purpose: "'Oh! Ahab,' cried Starbuck, 'not too late is it, even now, the third day, to desist. See! Moby Dick seeks the not. It is thou, thou, that madly seekest him'" (chapter 135, "The Chase—Third Day"). And, yet one wishes to say to Starbuck, "Hit Ahab over the head; do something, and take control!" Why does Starbuck bear it all? Partly, the answer is the chase itself, the quickness of the pace does not easily allow for reflective thought to flow into deed. Starbuck should have been prepared for the deed; but instead his past hesitancy now only allows the force of speech, useless on Ahab.

These final chapters, at least from 119 until the end, give us a mixture of Ahab's unwavering vow to remain true to his own need to kill Moby Dick and to have the crew follow him—"All your oaths to hunt the White Whale are as binding as mine" (chapter 119, "The Candles")—together with a strange showing of his more human side. Thus, Ahab befriends the former happy cabin-boy, Pip, who was almost drowned and whose loss of reason now seems strangely inviting. The strange bond between the two does not go unnoticed by the crew, as is clear from chapter 125, "Log and Line." This log is, at times, towed to measure the ship's speed and it is tied to a line. A Tahitian sailor, Manxman, has managed to bring the log back on board the boat, when the line snaps.

Ahab says that he can repair it as he repairs all things on the ship. Manxman says nothing in response, but then noticing Ahab taking Pip back to his cabin, mutters, "There goes two, daft ones now … One daft with strength, the other daft with weakness." Then too, Manxman is not taken in by Ahab; the line needs to be replaced and he will ask Stubb for a new one.

Thus, rather than being a linear narrative moving us straight ahead in pursuit of the great white whale, Moby Dick, these nineteen chapters, from 117, "The Whale Watch" to 135, "The Chase—Third Day," while peppered as usual with numerous reflections and allusions, reveal over and over Ahab's cutting insights into his own evil. But each insight, each opening for change is turned aside, and Ahab thus hardens his vow to remain true to his mad search after Moby Dick. For example, consider another encounter between Ahab and the carpenter. In my chapter 127, the chapter heading is simply, "The Deck," but I have provide a dramatic parenthetical introduction to the scene. The carpenter has recently fixed Ahab's leg, but now Ahab sees him altering Queequeg's coffin into a life-raft, since the regular one has been lost. Ahab notices him at work, and says, "Thou are as unprincipled as the gods, and as much of a jack-of-all-trades." The carpenter replies, "But I do not mean anything, sir. I do as I do." But that coffin will save Ishmael's life..

A Conclusion for Bildad and Peleg

> "The ship! The hearse!—the second hearse!" "cried Ahab from the boat"; "its wood could only be American!". (chapter 135, "The Chase—Third Day")

I have written a fast three-day chase after Moby Dick ends in disaster. Chapters 133, 134, and 135 continue with the simple sign-post character that has been typical of most of the previous chapters and thus "The Chase—First Day," "The Chase—Second Day," and finally, "The Chase—Third Day." In substance, these three concluding chapters are filled with swift adventure, deep cries of warning and pleading from Starbuck to give up the chase, and an unbelievable concurrence of Fedallah's predictions about Ahab's death.

First, let us consider the adventure. It is a mixture of an almost suicidal attempt to kill Moby Dick with an almost equal attempt to render the action believable by reminding us that, after all, this madness is part of the adventure of whaling. In the first day of the chase, Ahab is thrown from his boat by Moby Dick, and, while in the water, he shouts to the Pequod to drive itself between him and the whale. This is done successfully, and Ahab and his crew are then rescued by Stubb's boat. Ahab, now on Stubb's boat, does not go back to rest on the Pequod, but rather continues to chase Moby Dick with a double crew at the oars. I inform my readers not to wonder, "It is often the case that when a boat is stove, its crew, being picked up by another boat, help that second boat; and the chase is continued with what is called double-banked oars."

Still, Ahab almost drowned with all his crew, and yet the search to kill Moby Dick continues, even though the whale has sped away. But, again, we are not to wonder too much at the continuation of the search: "For such is the wonderful skill, prescience of experience, and invincible confidence … among the Nantucket commanders," that they can foretell where and when a whale they have chased will reappear. But, in fact, the Pequod first follows a false lead; and, then regains its direction, when Moby Dick breaches and the chase begins again: "They were one man, not thirty."

Having been rescued by the Pequod, the whaling boats are now lowered again. Ahab's boat is in the center, and he tells his crew that "he would take the whale head-and-head"—that is, pull straight up to his forehead. Once again, we are not to be too surprised, for it is "a not uncommon thing." Nevertheless, the maneuver is disastrous to all three boats—first Stubb and Flask, together with their crew are thrown into the water, and then Ahab's own boat is dashed by the great whale. Ahab's ivory leg is snapped off, and he boards the Pequod leaning on Starbuck, "Aye, aye, Starbuck, 'tis sweet to lean sometimes." All seem to be safe, but then Ahab asks to see Fedallah, who cannot be found, and Stubb recalls that he saw Fedallah caught among Ahab's own line, which was tied to his special harpoon.

Still, Ahab will not give up, "I'll slay him yet!" Starbuck again tries to turn Ahab away from his monomania, and this time in a rather long speech, ending, "Oh, oh,—Impiety and blasphemy to hunt him more." To which Ahab responds that, although he feels strangely close to Starbuck ever since he saw the longing for shore and home in his eyes, that he wishes Starbuck to "be the front of thy face to me as the palm of this hand—a lipless, unfeatured blank." Ahab sees himself as Fate's lieutenant, doing what must be done, and Starbuck is not to question Fate.

And now to conclude this adventure, we have the last day of the search. Again, Starbuck reflects that his obedience to Ahab is a disobedience to God, and there are tears: "Oh, my captain, my captain!—noble heart—go not—go not!—see, it's a brave man that weeps; how great the agony of the persuasion then!" But Ahab orders the boat lowered and tosses the mate's arm from him.

The action continues, and now there is no suggestion that, after all, it can be thought of as not too uncommon. True, sharks follow, and, while this is not uncommon, they follow only Ahab's boats, and in great numbers, chewing the oars. And, then, the chase is so mixed with portents and omens that no one can hint it is ordinary, even in whaling. Thus, a shout goes up, and Moby Dick rises with the body of Fedallah (the Parsee) lashed around his hump with the very harpoon lines he threw at the whale. Ahab now sees the first part of Fedallah's prediction come true, namely, that Ahab could not die until he first sees Fedallah precede him in a hearse not made from nature. But Fedallah had also foretold that Ahab could not die until he saw a second hearse made of wood from America, and finally, that he could only die from hemp. It will all come true.

Moby Dick, perhaps sensing that the Pequod itself is the source of all the evil pursuing him, aims his body at the ship: "Retribution, swift vengeance, eternal malice were in his whole aspect." The white whale shatters the Pequod into splinters, all on board fall where they have been working. Ahab looks at his sinking ship from his boat and cries: "'The ship! The hearse!—the second hearse!' … its wood could only be American!" Still, seeing the second part of the Parsee's prediction come true, Ahab continues his pursuit. He throws a harpoon at Moby Dick, the line runs foul, and, as he is trying to clear it, the hemp line rebounds about his neck—"he was shot out of the boat, ere the crew knew he was gone." Every aspect of Fedallah's prediction comes true.

All on board the Pequod die as do all those on the small boats chasing Moby Dick—except one, Ishmael. Ishmael had been given a place on Ahab's boat, and he alone manages to rise from the whirling waters. He spots Queequeg's coffin made into a life-buoy, and he is able to hold on to it—the sharks gliding by "as if with padlocks on their mouths." On the second day, he is rescued by the Rachel, which was still looking for their captain's lost son, but "only found another orphan."

CHAPTER 12

Conclusions, the Unity of *Moby-Dick*, and a Critical Reflection

Abstract If the central theme of Melville's great classic was Ahab's fixation on Moby Dick, Melville could have written a book less than half the size as the one which he wrote, and, in that half, there would be plenty of space for reflections on monomania and its evils. No, the unity of Melville's book comes, first, from the way the numerous literary, philosophical, and religious reflections are rooted in those magnificent beings, whales, and in the men and ships that pursue them, and, second, in the way these reflections illuminate our own lives.

The ambiguity of this pursuit is an aspect of those reflections. Piety and business, danger and bravery, excitement and discipline are all part of the hunt, a hunt that lasts three years, years that define and fix for all on board both the clarity and ambiguity of all human existence. The mates and harpooners know their duties and they are willing to risk their lives as part of their sea adventure, and, the crew in general is willing to follow orders. But then a captain, the lord and master of the ship, has the power to imbue a distinct character to the ship and its crew, a character that may or may not be decisive during the voyage.

All of the above must be kept in mind as we prepare for a true interpretation. Allegories abound in the text itself, and these must be given their due before one attempts a true interpretation. Further, whatever meanings one gives to the great white whale, Moby Dick, we must keep in mind

J. S. Catalano, *In Pursuit of Moby-Dick*,
https://doi.org/10.1007/978-3-031-40357-6_12

Melville's warning that only a person ignorant of whaling would interpret "Moby Dick as a monstrous fable, or still worse and more detestable, a hideous and intolerable allegory."

Keywords Epilogue • The Specksynder • Gam • The Great Heidelburgh Tun • Ambergris • The Try-Works • A Bower in the Aracides

"*The drama's done*," I have my Ishmael write in the Epilogue. True, but what precisely was the drama? In general, we know the answer. Let us recall how strangely it all begins. We are told that the "Etymology" has been supplied by a late consumptive usher to a grammar school—that is to say, by an underpaid part-time teacher—and that the eighty-odd "Extracts" have been supplies by a sub-sub-librarian—is the poor soul paid at all?—and that Ishmael himself is only the commentator of this "poor devil."

It is tempting to pass over these parenthetical introductory remarks as well as the two preliminary sections. And, the editors of the first English edition of my book, which was published by Bentley in three handsome volumes, didn't quite know what to do with them, and put them at the end of the tale. They also decided to leave out the "Epilogue" and, as I mentioned earlier, they did some substantial editing of material they thought offensive. But, to repeat, I, Melville, handed the American publishers of Harper and Brothers a complete, properly organized manuscript. Nevertheless, I did change the title; in the English edition, the book was titled *The Whale*; but I now named his work, *Moby-Dick; or, The Whale*. What were my reasons? Well, I wanted a complete copy—that most of all—and the original title, "The Whale," might appear to put my book in competition with another work recently published by Harper. Indeed, the substance of the book is largely about whales in general.

But, to return to the parenthetical remarks that introduce the "Etymology" and "The Extracts," their strangeness becomes a haunting overlay on the entire narrative. It continues through the very first chapter, "Loomings," and then re-appears periodically throughout the narrative. It is not of the same quality throughout. Nevertheless, if my readers are to understand the multileveled unity of my book, an attempt must be made, if not to fix this strangeness, at least to indicate its presence and it possible functions.

Again, if one glances over the 135 chapters, all the headings are easily understandable on sight, except seven or eight, for example, chapter 31 is "Queen Mab," that of 33, "The Specksynder," of 53, "The Gam," of 77, "The Great Heidelburgh Tun," of 92, "Ambergris," of 96, "The Try-Works," and the strangest of all, the title of chapter 102 is "A Bower in the Aracides." One may add or subtract from these depending upon one's literary background, and thus "Queen Mab," might have for you the same clear sign-post character as the previous chapter 30, "The Pipe." But even where the meanings do not shine from the titles themselves, most are explained in the chapters; and they reveal themselves to be either physical aspects of American whaling ships or aspects of whaling itself, as is clear from chapters 33, "The Specksynder," and 92, "Ambergris."

The one great exception to all the clarity in the titles is again that remarkable chapter, 102, "A Bower in the Aracides." The title is weird and the content weirder. But I see it connecting directly with that misty first chapter "Loomings," and, as I will explain, with Ishmael's remarkable escape. Once again, I recommend that we not be ashamed to note the obvious. Why the title, "Loomings," when immediately thereafter we then have a long series of very obvious chapter headings such as "The Carpet Bag," "The Spouter Inn," "The Counterpane (true, archaic for quilt, but still clear), "Breakfast," and "The Street." And why do we have to enter this cloud, "Loomings," in the very beginning of the tale, with the strange invitation, "Call me Ishmael?" Why did Melville not simply begin his tale, "Going Whaling," a title that would fit nicely with all the following; and why did Ishmael (if that is his name) not simply say, "My name is Ishmael"?

The first chapter is rightly titled, "Loomings," because not only is Ishmael's real name to remain in the mist, but a layer of the book itself. As we read (and re-read) this first chapter, we begin to realize that the entire narrative to be unfolded to us shares in the ambiguity of Ishmael's real name. For this tale is about water, its majesty and mystery, what lives below and floats above, about water, the source of all life, drawing us back into itself, and thus, in this same first chapter we read:

> And deeper the meaning of that story of Narcissus, who because he could not grasp the tormenting, mild image he saw in the fountain, plunged into it and was drowned. But that same mage, we ourselves see in all rivers and oceans. It is the image of the ungraspable phantom of life; and this is the key to all.

Of course, this book is filled with such reflections, and we must be careful not to overdo the significance of any particular one; for indeed, many are cited as the "key" to the narrative. On the other hand, this appears very early, and the entire chapter shares in the "ungraspable phantom of life." Indeed, the entire book shares in the twin drama of life and the phantom of life.

Ishmael wants to go whaling, and finally he does tell us that the main reason for his desire "was the overwhelming idea of the great whale himself." But this comes after a long reflection on fate and providence, a reflection that almost hints at Ahab's own thoughts: "And, doubtless, my going on this whaling voyage, formed part of the grand programme of Providence that was drawn up a long time ago." On reading this chapter, one might say to oneself, "Ishmael, for the sake of your tale, I grant you this appeal to providence; but then you detail for us sufficient reasons to account for your decision apart from any such appeal to the 'Fates.'"

And yet, those appeals to fate stand in the text. I, Melville, did not cross them out and hand the printers a simpler reading. For, indeed, not only this first chapter but the entire book is clear and misty, bright with enlightened reason, and dark with portents and omens. And, then, there are predictions of what is to come. The chapter begins with Ishmael's desire to go whaling and yet it concludes with a reflection on Moby Dick, a whale not be mentioned until chapter 36, "The Quarter-Deck," more than a quarter of the way into the narrative. But, here, in the first chapter, we read: "two by two there floated into my inmost soul, endless processions of the whale, and, midmost of them all, one grand hooded phantom, like a snow hill in the air."

Obviously, Ishmael knows how the adventure is going to end, and there is nothing remarkable in hinting at this ending this early in the text. But none of this explains the mixture of an appeal to fate and reason, an appeal that is to continue until the very end, with the impossible detailed description of Fedallah's prediction of Ahab's death. Thus, recall that immediately after the first chapter, "Loomings" mists, omens, and portents arise again in the next chapter, "The Carpet-Bag," Ishmael looks at the sign of the inn he is about to enter and he sees:

> a swinging sign over the door with a white painting upon it, faintly representing a tall straight yet of misty spray, and these words underneath—"The Spouter-Inn:—Peter Coffin." Coffin?—Spouter?—rather ominous.

Still, omens and portents notwithstanding, he enters; and the tale continues with omens and portents—*none of which alter anyone's behavior.* But, then, why include them? Well, sailors are superstitious. Yes, but that does not explain the extent of the constant reappearance of these omens. I suspect that we are concerned with something deeper than telling a tale from memory. I think the tale itself is being modified for us, beyond any ordinary alteration that comes from the attempt to remember and retell things past.

Perhaps he is not happy with how it turned out. Perhaps, also, he is looking for excuses for his own weakness before the commanding presence of Ahab and for the weakness of the crew, particularly, the chief mate, Starbuck. True, he admits his weakness. Chapter 41, "Moby Dick," begins, "I, Ishmael, was one of that crew; my shouts had gone up with the rest; my oath had been welded with theirs." And yet this chapter is filled with extraordinary powers attributed both to the White Whale and to Ahab's monomania and evil: "This is much; yet Ahab's larger, darker, deeper part remains unhinted." "Unhinted," and, yet, Ishmael did a very good job detailing for us this evil, both in roots and in its appearance on the Pequod. Recall, for example, the descriptions given to us in chapters 33, "The Specksynder," 41, "Moby Dick," and 44, "The Chart," as given above in my chapters, 5 and 6.

On the other hand, I wish to be very careful about what I attribute to Ishmael 2, since Ishmael 1, the main character along with Ahab, and the omniscient narrator is himself highly reflective, and little passes his notice. And yet, I repeat my earlier reflection about chapter 41, "The Whiteness of the Whale," namely, that it is, for me, not a consistent part of Ishmael 1; it is too phony. I see here the hand of Ishmael 2; namely, an Ishmael now at odds with himself, no longer attempting a faithful narration—with all the ambiguity that that implies—but now an Ishmael consciously altering the tale, or, at least, providing us with doubts, doubts that might excuse his own weakness and that of the crew on that first voyage with the Pequod. For the present, let us not search deeper for the reason for those doubts, but, once again, attempt to connect the first chapter, "Loomings" with chapter 102, "A Bower in the Aracides." Are we, then, expected to accept chapter 102 as part of the original drama, a drama that details for us the *truth* of the greatness of whales and whaling? It seems, rather, a different kind of reflection:

> The skeleton dimensions I shall now proceed to set down are copied verbatim from my right arm, where I had them tattooed; as in my wild wandering at that period, there was no other secure way of preserving such valuable statics.

We would be completely wrong, however, if we concluded from this humorous presentation that the basic dimensions given about the sperm whale in the tale as a whole are false. Indeed, the narrative forces us to conclude the contrary. What we begin to realize as we read is that we are given two tales both told by Ishmael, but not in the same literary voice. Ishmael 1 is recounting to us a tale in a more or less straight-forward narrating voice, albeit with numerous reflections, because he is a reflective person, with a long history of whaling. He is a seasoned sailor, before he takes time out to write this narrative about his first whaling experience. Ishmael 2 is this same Ishmael 1, but now, for some reason, he is consciously altering the adventure for us, even as he is writing. *Before proceeding any further let me repeat my interpretation for the clue to the entire narrative, namely, it is a twice-told tale, and there are entire chapters that seem to fit into another narrative that is being hinted at even as he is writing his main narrative.* Thus, we have two accounts of the general great size and features of the sperm whale as well as every aspect of pursuing and killing these leviathans of the sea, two accounts that are at odds with each other. For example, from chapters 55 to 90, we have an almost continual discussion of whales and whaling, and to these we must many others so that almost two-thirds of the book does not concern Ahab in his mad pursuit after Moby Dick. And, yet, chapter 102 exists, a chapter of considerable length—it will not disappear for being embarrassing. I see it as the clear entrance, or rather, re-entrance of Ishmael 2 into the narrative warning us that the ending of the narrative did not really happen as given and as written by him to satisfy those who sent Ahab to sea again, perhaps too early.

Did Ahab come home wounded and would those who allowed him to go rather he be remembered as having died at sea, having killed his whale even if he was also killed by it? Great writing delivers truth, as well as the gift of freedom to its readers. Did the crew throw Ahab overboard as the companions of Job threw him, apparently with God's blessing and with Job's own approval? I do not find this difficult to image as part of the tale. Ahab, realizing that he cannot let go of hatred of Moby Dick, but not wanting to meaninglessly endanger the lives of his crew, takes one boat with Fedallah and his Chinese crew to chase the whale. All these die, and the Pequod sails safely away, landing on a remote island, while Ishmael many years later is asked by the crew to write the tale we are given.

Or, perhaps all return with Ahab wounded, and whisked secretly away to live at home, with perhaps Pip, his child and his "resigned" wife, fading in and out, but basically content, believing that he killed his whale, as the Pequod is refitted with Starbuck as captain, and all the other crew mates, except Queequeg who has returned to be king of his Island, and with Fedallah and his crew sent off elsewhere. But, yes, the ending given is the one Peleg and Bildad sort of wanted, with Ahab dying a hero's death of sorts. I think that they wanted to give Ahab a marble homage alongside those others in Father Mapple's Chapel, but, of course, there is no more room.

My tale invites the reader to write a new ending, your own, and to share in the mystery of making a book. What is a book? A book is marks and spaces on paper or on a screen. *Moby-Dick; or, The Whale* is composed of alphabetic marks, spaces, and punctuations arranged to deliver the message within the book. Queequeg looks at similar marks about whales, counts their pages, and nods his head in agreement when Ishmael hints at their meanings. Throughout the narrative we are informed about the make-up of books—for example, the early character of the printed word, "black-letter," from the heavy type used. And, there is that long comparison in chapter 32, "Cetology," which many critics simply see as Melville's debunking of classifications; but which is, nevertheless, fairly accurate and, as it stands, a lengthy discussion of books as books and their relation to their content.

Indeed, the mystery of every book is the magic of how marks and spaces deliver meaning to the literate. And that mystery is at least as deep as the oceans themselves and as great as the leviathans which swim and breach within these oceans. The mystery about how a narrative works to deliver its message, and the truth of that message, follow us throughout this tale. For what that poor usher dusts are not merely long-lived objects that will endure after he dies; but Ideas housed within its bindings and pages.

Ishmael, who knows his Plato, is not unaware of the claim that our notions of good and evil are housed in a world apart from this world. But, an author, if true to his or her discipline is more of an Aristotelian, giving human efforts its rightful due of crafting marks into meanings, and, then, perhaps, into ideas themselves. In *Moby-Dick; or, The Whale*, we thus read not merely an adventure of whaling but an adventure of the words delivering that adventure to us, the adventure of the whale itself, the word whale, and the books that speak to us of all this, books created by us, the eyes, ears, fingers, and all that give us books to read.

Still, my *Moby-Dick* does concern the broadest issues of good and evil, both within our own lives and within the social order. Perhaps, after attempting to give due credit to the richness of the text itself—a task not frequently done—one can offer an implicit analogy running within the tale. I will give one. It concerns the mistaken belief that doing something good is the same as eliminating an evil. Indeed, at times, this is what we have to do: to stop someone in the act of committing murder is to save a life. But as a social directive, it is deeply misguided; it puts us in the constant need to find enemies to defeat.

This social attitude can be seen as a re-birth of a Christian heresy named, "Manichaeism." Manichaeism, in its Persian and Roman roots during the third and fourth centuries A.D., held that history reflects the eternal battle between two supernatural forces—Good and Evil. In the attitude of Manichaeism, the elimination of an evil is the same as the creation of a good. And this is how Ahab chose to see his battle with Moby Dick. The white whale became for him the embodiment of all the evils of the world, and its death meant that good could enter the world again. In this way, Ahab could see himself beyond the ordinary, a man chosen to duel the devil.

Today, Manichaeism continues. We need enemies. We act as if we want to live in a world without this evil, but we need our wars, for these define our good. And, if we should indeed turn to constructing good, we are tame and hesitant, afraid to assert the creative aspect of our humanity. The failure of Starbuck and the other mates to assert themselves against Ahab is our failure. And, if there is to be another ending of the tale of Pequod, we will have to awaken ourselves from our slumber. But those in power are not stupid; they keep us busy earning our bread, while they forge unemployment and poverty as the milieu of our existence, making it as difficult as possible for us to wonder, even in a world filled with wonder, the wonder of whales and of books about whales.

Critical References

Whether Eric Williams is right that D. H. Lawrence revived Herman Melville's *Moby-Dick* he recommends that we all read the book, and I have done such. With this in mind, let all of those who have written about this great book advance to recommend its reading and the commitment that reading implies. As Toni Morrison writes in "Playing in the Dark: Whiteness and the Literary Imagination" (Harvard University Press, 1992), all dedicated reading is a *becoming*, "imagining is not merely looking or looking at; nor is taking oneself intact into the other. It is for the purposes of the work, *becoming*" (4, author's italics).

Thus, let us *become* one with Ishmael as he leaves home heading for a voyage in which he will become the *Ishmael* of that voyage: "Call me Ishmael," we read in wonder. Are not these opening three words as strange as the haunting title of this first chapter, "Loomings?" The name "Ishmael" has Biblical references, but that is not what I see. I see three words a universe apart from the simple claim, "My name is Ishmael."

Kerry McSweeney's *Moby-Dick: Ishmael's Mighty Book* (Boston: Twayne, 1986) offers us Ishmael as both a narrator and a central character. McSweeney writes, "His [Ishmael's] vision is different in one important particular: his awareness of the epistemological problem of the subjectivity of symbolic perception and of the consequent relativity of symbolic meaning." As a philosopher and one who has read *Moby-Dick* on and off for the past thirty-five years or so, I find this far denser than Melville's book itself; but its flavor is unfortunately characteristic of many critics.

The eminent critic Harold Bloom in *Herman Melville's Moby-Dick* (New York, Infobase, 2007) asserts that *Moby-Dick* centers on Ahab's monomaniacal pursuit of the whale that chewed off his leg; he "argues for Captain Ahab as the American Promethean hero, and as a Gnostic visionary, rather like Herman Melville himself." I am not interested in the life of Herman Melville, but only in the book, *Moby-Dick*, itself, and here, Bloom and others who see the book center only on Ahab are wrong. Even forgetting Melville's own words about his coming book, this tale about Ahab, as I will demonstrate, fills, at best, only about a third of the book. Bloom gives us blubber, no poetry, and thus error instead of truth.

Then too, Janez Stanonik's *Moby-Dick: Myth and the Symbol* (Yugoslavia, 1962), and H. B. Kulkarni's *Moby-Dick: A Hindu Avatar. A Study of Hindu Myth and Thought in Moby-Dick* (Utah, 1970), center their studies too much on Ahab's search for the whale, Moby Dick. But, aside from these, there is also a very important and perceptive recent study that interprets Melville's epic as being swallowed by the philosophy therein. Andrew Delbanco, in *Melville: His World and Work* (Knopf, 2005), writes:

> When *Moby-Dick* finally appeared at the end of 1851, it did not restore its author's reputation. It was really two books—a going-to-sea story, and a

daring metaphysical adventure—and when the second swallowed the first, it swallowed it whole, leaving it intact like Jonah in the belly of the beast.

And, later in the same book, Delbanco writes:

The many passages in *Moby-Dick* in which whales are listed, sorted, described, defined.... These cetological chapters retard the pace of the narrative, and many readers prefer to skip them in order to get on with the great chase; but it is here that Melville makes his case, with tongue in cheek, against all forms of classification—including the racial form.

I disagree on both accounts. Granting that the metaphysical adventure swallows the going-to-sea story, I claim that the going-to-sea story then returns, not to swallow, but to incorporate, the metaphysical adventure so that finally there is one adventure, thoroughly American and thoroughly imbued with literary, philosophical, and religious allusions, a tale in which truth and poetry come from blubber and in which the reader is clearly informed that this is the very tale itself.

Consider also the comments of the perceptive Louis Mumford in *Herman Melville* (New York: Literary Guild of America, 1929):

But *Moby-Dick*, admirable as it is as a narrative of maritime adventure, is far more than that: it is, fundamentally, a parable on the mystery of evil and the accidental malice of the universe. The white whale stands for the brute energies of existence, blind, fatal, overpowering, while Ahab is the spirit of man, small and feeble, but purposive, that pits its puniness against this might, and its purpose against the blank senselessness of power.... All our lesser literature is an evasion of the white whale.

I might quibble about this somewhat noble depiction of Ahab; but there is no need, for I both think that Mumford is right and wrong. He is right in the sense that *Moby-Dick* is more than an adventure story; but he is wrong in focusing on the parable aspect, for "this more" comes from the "blubber," that is, from the literal tale of the whaling story itself. The story itself is not a parable. The tale is the very tale we read: Everything is on the surface of the book; the Pacific Ocean may be calm as a lake or its waves may be as tall as a sky-scrapper, but it is that surface that affects the whaling ship, the crew, and whether it can catch whales. Melville does not hide his reflections; he does hide his emphasis on whaling as a business; he does not hide the greatness of the whales themselves. They are all there, and, as we will see, for the most part, in the same chapter.

Indeed, the expansiveness of the book has not gone unnoticed; for example, it is termed an "epic" rather than a novel proper. Thus, Jean-Paul Sartre, who I admire as a philosopher, writes about *Moby-Dick*:

> There is an idea of hatred just as there is an idea of whiteness or of the whale hunt, and this ideal involves the whole man, the whole human condition. From now on the novelist's technique seems to Melville to be insufficient to catch this idea. All means are going to seem legitimate to him: sermons, courtroom oratory, theatrical dialogues, interior monologues, real or seeming erudition, the epic—the epic above all. (*Herman Melville's Moby Dick contained in Selected Prose Writings of Jean-Paul Sartre* (Northwestern: Northwestern University Press, 1974)

Even in this short study, Sartre correctly makes the observation that all the forms of discourse are on the surface of the text, and frequently with a good deal of humor, as noted by other critics. Thus, toward the end of the book, the chapter headings include quasi-play directions. Sartre thinks that, as Melville was writing, he gradually understood that his tale could not be forged as in his earlier works. And, while it is true that a writer frequently discovers what is being invented, I think that here Sartre is wrong: The epic, if such it be, is being born from the opening lines, "Call me Ishmael." What a strange and friendly invitation to a reader! Still, Shakespeare is, indeed in this tale, not hidden, but in full view with Cervantes, Job, Jonah, Zarathustra, Plato, and a host of other notables.

Thus, even if we grant with the American poet, Charles Olson, who, in *Call Me Ishmael* (San Francisco, City Lights, 1947; John Hopkins, 97), made the case for a strong Shakespearean influence on Melville, it is true that Melville discovered a large-print edition of Shakespeare (his eyes were bad) as he was planning to write *Moby-Dick*, and Olson noted that Melville had written notes on the last pages of the set. But, I am not the only one to doubt that this indeed means that Melville had thereby decided to alter the book toward *King Lear*, for example. On the contrary (putting aside his attempt to view the book in terms of myth, space, and tragedy, and forgoing here the many good aspects of his study), I think it overly convoluted. Still, surprisingly, he notes Melville's humor in regard to Shakespeare, and it is the same thought that occurs to me. I frequently have the impression that Melville is making fun of time and a nation that drew its inspiration from kings and nobles. And, even as he is serious, he has fun, for example, with philosophy, as, when considering the spouting of whales, he writes in chapter 85, "The Fountain":

> And I am convinced that from the heads of all ponderous profound beings, such as Plato, Pyrrho, the Devil, Jupiter, Dante, and so on, there always goes up a certain semi-visible steam, while in the act of thinking deep thoughts. While composing a little treatise on Eternity, I had the curiosity to place a mirror before me; and ere long saw reflected there, a curious involved worming and undulation in the atmosphere over my head.

Yes, *Moby-Dick* is big, and even the most sympathetic of readers are inclined to apologize for its length. Thus, even Hershel Parker, who hints at the greatness of Melville's book, claims that the details of the whaling scenes are needed as "ballast" for the book. A third of the book for ballast? But Parker's recent and great two-volume biography (John Hopkins University Press, 1996; 2002) does not attempt to offer us an interpretation of *Moby-Dick*, although, in general, he prepares us for a proper reading.

Indeed, the early reviews complained that the book was long-winded; for the most part, the critics complained that Melville should have just gotten along with the tale of Ahab's pursuit of Moby Dick. And, unfortunately, it was as a mere teller of sea adventures that, even his (supposedly) close friend, Evert Duyckinck, saw him. Parker, in his second volume notes about Duyckinck: "His view of *Moby-Dick?* Melville had begun as a sea-writer and in Duyckinck's mind Melville stayed a sea-writer." Interestingly, Bloom, in his *Bloom's Notes on Moby-Dick*, includes a short excerpt of Duyckinck's review with the comment, "Duyckinck presents a mixed but on the whole favorable view of the work." But, for Parker, Duyckinck missed the entire point of Melville's classic. I agree, and I add the same is true of Harold Bloom!

Sadly, Melville's masterpiece was, for the most part, rejected by its original readers. Richard Bentley, the publisher of the first British edition of only 500 copies lost money. And, of the 3000 copies of the American edition, a little more than half were sold. Furthermore, "when in December 1853, the unsold copies burned up in a fire in the publisher's warehouse, few noticed and fewer cared" (Delbanco). "Melville found it harder, day by day, to see himself as the author of a great Shakespearean whaling book" (Parker).

The poor sales show that most readers were confused by the book, but they might have been helped by a competent fleet of critics, but, for the most part, these were lacking in America in the 1850s, and most critical eyes were those of English reviewers. Some of these were actually very good, but Melville unfortunately did not know about them, and they were not such as to interest the average reader. Little by little, however, America grew in self-confidence, and slowly, gathering strength, a tidal wave of praise gathered, both here and indeed in England, praising the book as one of America's greatest literary achievement, if not, indeed, its greatest, justifying Melville's own judgment about his achievement. And thus, I must not give the impression that I am against critics, for without them Melville's great book might have been lost to us, and to me in particular.

In re-reading Tony Hillerman's *A Thief of Time*, I noticed this review by Robert B. Parker on the inside flab:

> In *A Thief of Time*, Tony Hillerman does what only the very best of writers can do. He gives us a compelling story, richly detailed, about people that matter to us, in lucid prose, and, at the same time, he makes a kind of magic which touches us in ways we don't totally understand.

I have made an effort to understand some of the ways *Moby-Dick; or, The Whale* touches us. Still, the second part of this title reminds me of the interesting history of the publication of this book. We read these words from Milton's *Paradise Lost* are on the title page of the true first edition of *Moby-Dick*; and, further, Melville's title of that first British edition is simply, *The Whale.*

> There Leviathan
> Hugest of living creatures, in the deep
> Stretch'd like a promontory sleeps or swims,
> And seems a moving land; and at his gills
> Draws in, and at his breath spouts out a sea

As I collector, I have always wanted to own that book, but it was beyond my means, for only 500 copies were printed in England by Richard Bentley in October 1851. Its production was strange and somewhat contradictory; for, while it came out in three impressive-looking volumes, it was incomplete. Bentley trimmed Melville's classic by thirty-five passages that he thought offensive. He also put the two preliminary sections of the book, "Etymology" and "Extracts," at the end; and, he left out the last page, "Epilogue," describing Ishmael's escape from the doomed ship, Pequod, with the result that the British readers and critics wondered how this tale of more than 600 pages could have been narrated.

A short time after its publication in England, Melville delivered a complete and properly organized manuscript to Fletcher Harper (one of the Harper brothers) of Harper & Brothers in America. There was one important change. Harper had recently published Henry T. Cheever's, *The Whale and His Captors*, and, perhaps anticipating complaints that he was duplicating Cheever's efforts, Melville decided to change the title of his book (perhaps, in an effort to make it seem more specific), and thus, almost *as a last minute decision*, Melville altered the title from *The Whale*, to *Moby-Dick; or, The Whale*. Notice the semi-colon; it is not meant to be redundant. "*The Whale*" in the title does not merely refer back to the one whale, Moby Dick, but it means all whales, and this is evident from the book itself. Nevertheless, although this is the true complete title of the book, clear for all to see in the first American edition, it is seldom repeated. It is more convenient to refer simply to *Moby-Dick*, and I will frequently follow that custom, although I cannot but wonder if our reading would have been guided more properly by the original title of the British edition—***The Whale***. Whales! Who could have imagined that such beings exist on our planet? Impossibly huge, mammals, not fish, they swim the oceans as we walk our highways, and their existence makes the book itself not allegory but wondrous truth.

And thus we return to *becoming* what we read, thinking of life and death as if we were on a whaling ship, as Melville writes in chapter 60, "And if you be a philosopher, though seated in the whale-boat, you would not at heart feel one whit more of terror, though seated before your evening fire with a poker, and not a harpoon, by your side."

And now also, a deeper meaning of our book unveils itself in the sense of Morrison's full title, "Whiteness and the Literary Imagination," as the blackness of Pip, the cabin boy, who takes care of Ahab, and Queequeg, the primary harpooner, black, with tattoos that speak of white oppression.

Index

J. S. Catalano, *In Pursuit of Moby-Dick*,
https://doi.org/10.1007/978-3-031-40357-6

Kolibri 360 GmbH
Ferdinand-Jühlke-Straße 7
99095 Erfurt, Deutschland
produktsicherheit@kolibri360.de